MAN'S PLACE IN NATURE

Some dozen years after Père Teilhard de Chardin had completed *The Phenomenon of Man*, he set out to describe as concisely and accurately as he could what exactly is represented by man within the structure of the cosmos, what part he has played historically, in what direction he must continue to evolve, and what conditions must be fulfilled if the rebound of evolution in which we are now living is successfully to attain its term.

A close but clearly presented argument carries the reader along the 'curve of complexity', introducing him to Teilhard's key-principle of 'complexity-consciousness', across the thresholds that introduce life, consciousness and reflection, and places within this framework the specially favoured off-shoot from the tree of life that evolved into man. Teilhard's mind was always directed to the future, to the increasing socialization of humanity and the problems raised by the conflict between totalization and personality. It is here that he did most to inspire hope, through his belief that man, if he chooses correctly, may fulfil himself by attaining the 'universal focus' of a Christified cosmos.

By the same author in Fount Paperbacks

THE PHENOMENON OF MAN
LE MILIEU DIVIN
LETTERS FROM A TRAVELLER
THE FUTURE OF MAN

*about Teilhard de Chardin
in Fount Paperbacks*

AN INTRODUCTION TO TEILHARD DE CHARDIN
by N. M. Wildiers

EVOLUTION: THE THEORY OF TEILHARD DE CHARDIN
by Bernard Delfgaauw

PIERRE TEILHARD DE CHARDIN

MAN'S PLACE IN NATURE

The Human Zoological Group

TRANSLATED BY RENÉ HAGUE

Collins
FOUNT PAPERBACKS

First published in the English language by
William Collins Sons & Co. Ltd., London,
and Harper & Row, Inc., New York, 1966
First issued in Fontana Books 1971
Fourth Impression March 1974
Reprinted in Fount Paperbacks, March 1977

© 1956 by Éditions Albin Michel
© in the English translation
William Collins Sons & Co. Ltd., London,
and Harper & Row, Inc., New York, 1966

Made and printed in Great Britain by
William Collins Sons & Co. Ltd., Glasgow

CONDITIONS OF SALE
This book is sold subjec to the condition
that it shall not, by way of trade or otherwise,
be lent, re-sold, hired out or otherwise circulated
without the publisher's prior consent in any form of
binding or cover other than that in which it is
published and without a similar condition
including this condition being imposed
on the subsequent purchaser

INTRODUCTION

Père Teilhard de Chardin is at last coming into his own, and his life purpose is being achieved. The obstacles in the way of the dissemination of his vision of the world have now been removed. But he was a unique figure who cut his way through what in some sense were virgin forests of the mind, and there has been some misinterpretation. The misinterpretation has come from both religious and scientific milieux. It could hardly be otherwise as in his person Teilhard concretised two pursuits of man which have ignored one another for centuries—religion and science—and both at his own peculiarly high level.

For many official followers of Christ, brought up in a tradition that has hardly changed since the great Aristotelian-Thomist synthesis of religion and science in the thirteenth century, a Jesuit priest had no business exploring the panoramas opened to us in the twentieth century by astronomy, physics, biology and the other sciences 'in Xto Jesu'. For some of his fellow-explorers into the nature of the physical universe, Teilhard had no business to go beyond the limits of what is experimentally verifiable. Misunderstanding was made worse by the prodigious quantity of knowledge now available to man, which involves us in more and more specialisation. So that if theologians who first read Père Teilhard were ignorant of modern physics or biology so some physicists were ignorant of the very terms, or possible justifications, of faith in precisely St. Paul's (and Teilhard's) sense of 'the substance of things hoped for, the evidence of things that do not appear.' The situation was hardly made easier by the fact that Teilhard had to forge a new language to express new concepts—not only 'Noosphere', for instance, but above all 'Omega Point', which seemed some sort of unverifiable and fanciful poetry to metaphysical agnostics when they first read *The Phenomenon of Man*. When that vision of the world in terms of cosmic evolution was followed by *Le Milieu Divin*, in which we could hear the voice of St. Ignatius of Loyola, St. John of the Cross and (a non-Jansenist) Pascal expressed in a terminology of twentieth-century man in full crisis of 'cosmogenesis', confusion—here and there—was even greater.

If this note, therefore, is to follow Père Teilhard's intentions, it needs to dwell on two points. First the avalanche of the revolution in which contemporary man is involved, and second, the nature of faith.

When some four hundred years ago Pascal wrote, 'The eternal silence of these infinite spaces frightens me,' his uneasiness was based on a scientific perception of the universe that was already wider than the static type universe of St. Thomas Aquinas and Dante. But Pascal's universe seems tiny indeed compared with the picture that the contemporary sciences have revealed to us. Astronomy is probing further and further into space every month, counting distances in its millions of light years. Physics have involved us in a view of matter and its structure which might seem to have knocked the bottom out of all our stable convictions. There are the possibilities of life on other planets and of creating life or obliterating it, which, though a subject of Science Fiction, are also real enough. We have the leaping developments of cybernetics as well as medicine. The analysis of the psyche is perhaps only in its infancy. It is approximately in these perspectives that Teilhard discusses man's place in nature. There is now a literature of anguish about man's situation that goes far beyond what was thinkable in Pascal's time when he made the *pari*. There is at least a background of fright behind

2

our neuroses if we are not in danger of depression and despair. Moreover the progress of science, once speedy, has now become so vertiginous that in one year we have more advances than were made in a century of earlier time. There seem few foreseeable limits to this onrush[1]. Our perspective of history has been completely changed. Man may need roots, but where are they?

Père Teilhard was more completely at ease than most men in this new and terrifying vision of the world. It was as though, when he regarded the construction of matter or the boundless expanses of space, he said from his Roman Breviary with the confidence of a child: *Caeli enarrant gloriam Dei et opus manuum ejus annuntiat firmamentum.* As the serenity of Teilhard's faith may seem to make him an anomaly when most men 'at the spearhead of evolution'—to use one of his favourite expressions—'hic et nunc' feel bound to take an agnostic view of 'things unseen', and at the same time (witness the popularity of his works) gives courage to the frightened, we may be pardoned if we dwell a little on this, our second point.

Teilhard's faith in 'things unseen' cannot be separated from his enthusiasm for the boldest and most advanced speculation regarding the scientifically verifiable or yet to be verified. For him being is unitary, the universe is 'convergent', yet, as his life showed, his faith was the driving force in what he did from the very beginning. Faith can neither be proved nor

[1]The efforts made by the *Rencontres Internationales* at Geneva to bring together experts in the sciences and other educated men, to discuss the new implications, have underlined some aspects of this point. A first class man in a specialised field lives in a world that transcends all frontiers, his colleagues are to be found in all the 'evolved' societies. The language barrier is a trifle compared with the barrier of ideas, viz. only his colleagues can understand him. The mass of people even in the most 'evolved' countries have no conception of the speed and immensity of this avalanche of scientific advance.

disproved by scientific exploration. But Teilhard's faith is a fact which is verifiable to all[1]. If the language in which he expressed his faith seems almost 'monstrously' different from that of the formal face with which Christianity has presented itself to many in its short history so far, the difference, after all, is no greater than the difference between the Aristotelian-Thomist, or pre-Copernican view of matter and the universe, and the dynamic evolutionary world which has now been revealed to us.

In other words Teilhard's faith was in no sense in a 'separate compartment' from his scientific views, and still less was it some sort of 'sickliness', in Nietzsche's sense, that sought for 'consolation' in the terrors of the night. The centre and meaning and point of the universe was Christ. The 'end' of it was the mystery of St. Paul's Pleroma. The motive force was love. It is a vision of being every bit as 'Christocentric' as the famous medieval synthesis. It is the credo expressed in terms of the now utterly different perspective of physical being. Teilhard's assent to this credo was not nominal or verbal. It was testified to in a living way by the life-long vows of poverty, chastity and obedience which he took when he became a member of the Jesuit order, and in which he saw his fulfilment.

In *Man's Place in Nature* Teilhard is less concerned with those aspects of being which are matters of faith than he was, for instance, in the second part of *The Phenomenon of Man*, in *Le Milieu Divin*, in *Hymn to the Universe* or in a great deal of his personal correspondence which has now been published. Here we have primarily the Teilhard who was a figure with

[1] I would refer the reader not only to Claude Cuénot but to *La Prière du Père Teilhard de Chardin* by Henri de Lubac (Arthème Fayard, 1964). Henri de Lubac is a fellow Jesuit of Teilhard's and, though somewhat younger, knew him well. He has done more than anybody else to clear up ambiguities about the 'orthodoxy' of Teilhard's faith and practice.

an aura, but seen too little, in the scientific corridors in Europe and America. Yet perhaps, even on this occasion, a further comment should be added for his fellows at the spearhead of 'man's evolution on earth' and for 'Christians' who are rarely as ardent as he was. There is no 'logical' sequence between the scientific statements made by Teilhard, and the statements that flow from his faith. There is no 'proving' the idea of an Omega Point from the present and future evolution of man. Teilhard's works constitute no sort of Summa. He was not a professional theologian or a philosopher. He produced no theory about the arts and their relation or absence of relation to the evolution of man or of science. He had not the faintest intention of becoming a solitary prophet regarding man's place in nature in the present dilemma. His vision of the Pleroma was a statement of the way in which he, imaginatively, as a man of intense faith, saw the assertions of the Christian revelation working out in the sort of cosmos science now knows we inhabit: a 'world' which he described with a torrent of love and confidence.

One final point. Teilhard was in the midst of writing *The Phenomenon of Man* in 1940, during his second period in Peking (he spent his war years in Japanese internment). The manuscript finally reached Rome for ecclesiastical censorship in 1944 and in the August of that year Teilhard learned that permission for publication had been refused. Teilhard could well be a hero for all vocational men who have worked under terms of censorship or misunderstanding. He persevered. The French of *Man's Place in Nature* was submitted to the ecclesiastical authorities in 1950. Teilhard was told that, as before, he had gone outside the purely scientific sphere.

This book will soon be approaching the twentieth anniversary of it's composition. Since John XXIII there has been a change of heart towards many things in Rome, for which

Teilhard always hoped. No man would have welcomed with greater joy than Teilhard the evolution of scientific exploration in fields from the question of entropy to the origin of life in these years since he died—suddenly in New York on Easter Sunday 1955.

BERNARD WALL

PREFACE

Père Teilhard de Chardin once related how, between the two allied notions of the genetic structure of fauna and the genetic structure of continents, he had come to see a third notion, that of the genetic structure of Mankind. From that time all his efforts were directed to building up an anthropogenesis, a science, that is, of man conceived as an extension of a science of life. It was a vast undertaking, but one fitted to so powerful a mind. He died, unhappily, while still in full intellectual vigour, before he could complete his task; but he left us its general plan, while various articles in which, in his own words, he had included all that was most valuable in his own experience and the essential core of his vision, give us a fully matured expression of his views on several fundamental points.

One aspect of anthropogenesis is studied in detail in this book, a classic aspect but treated by Teilhard in a new way, the problem of man's place in the structure of nature and the value he represents in it.

Père Teilhard gives us the fruit of his own personal reflections, and, in a magnificent picture, traces for us the "ascent" towards man in which lies the profound significance of cosmogenesis.

Life is not a chance combination of material elements, an accidental product of world history, but the form assumed by matter when it reaches a certain level of complexity. It introduces us to a new order, the biosphere, characterised by special properties. The biosphere is not to be envisaged as a purely spatial image, a mere envelope concentric with the lithosphere and the hydrosphere, a sort of framework within which life is enclosed, but rather as a structural layer of our planet, "a system in which we can see the connection that links together, within the same cosmic dynamism, Biology, Physics and Astronomy."

Life very soon manifests one of its fundamental tendencies, the tendency to ramify as it progresses. More fully and clearly than any other writer, Père Teilhard brings out the importance of the notion of the line of descent, or phylum, which is the true elementary unit of the biosphere. This latter resolves itself into a multitude of lines of descent; it presents a fibrous structure. Moreover, life never continues for long in the same direction; every line is sooner or later replaced, and to some extent, too, prolonged, by a lateral line, so that the fibrous structure of the biosphere appears, at the same time, as a scale-like structure.

At first sight, this bushy growth of life gives the impression of a diversity that defies analysis, of a profusion in which it is impossible to discover any natural order. On one of these multiple branches, following a mutation similar to all the others, man appears; and we might imagine that his slowly acquired superiority was no more than an accident of life.

Is this, however, a true picture of the phenomenon? In all this proliferation are there no specially favoured lines? To what extent, in any case, are we justified in introducing into such a problem the notion of value? These are the questions Père Teilhard answers.

Starting from a certain level of complexity, matter becomes "vitalised," and, on this plane, new qualities begin to emerge. Some, such as assimilation and reproduction, are to be found again in practically the same form, in the great series of Metazoan animals. On the other hand psychism, as early as the infra-human zones, brings in a hierarchical factor, a measure of the degree of vitalisation.

It is the intensity of psychism that distinguishes the two major lines of the metazoa, the arthropods and the vertebrates, by the development of instinct in the former and of intelligence in the latter.

Throughout the whole line of descent of the vertebrates, which is the only one that concerns us from the point of view of anthropogenesis, we see an increase of cerebralisation, from the fishes to the mammals. And among the latter, one group, from the same point of view, takes precedence of the others, the group of Primates. It represents a specially favoured axis of evolution. At the same time, within its various branches, this "effort" that life makes towards cerebralisation, sooner or later comes to a halt, the psychism failing to cross the threshold into reflection. Only in man does "consciousness snap the chain," and in man the highest aspiration of the phenomenon of life finds its fullest expression. Though there is no break in continuity with what precedes him, the arrival of man marks a completely new level, equal in importance to that constituted by the appearance of life, and which we may define as the establishment on our planet of a thinking sphere superimposed upon the biosphere: the noosphere.

It is in this that the vast effort of cerebralisation that started on the infant earth is to be fulfilled, in the direction of collective organisation or socialisation.

In the latter part of this book, Père Teilhard may well seem, indeed, to be writing rather as a philosopher than as a scientist, and many who have admired the palaeontologist's interpretation of the evolution of the living world may have some difficulty in following the writer in his vision of the future. No one, however, can fail to be impressed by the force and clarity of the thought, the intellectual mastery, of one of the greatest minds the world has known.

JEAN PIVETEAU

CONTENTS

Introduction by Bernard Wall   *page* 1

Preface by Jean Piveteau   7

Foreword   14

I. THE PLACE AND SIGNIFICANCE OF LIFE IN THE UNIVERSE. A SELF-INVOLUTING WORLD   17

    1. Physics and biology: the problem   17

    2. The basic proposition: different forms of arrangement of matter: "true" and "false" complexity   19

    3. The curve of "corpusculisation": life and complexity   21

    4. The mechanism of corpusculisation: the transit to life   26

    5. The dynamism of corpusculisation: the expansion of consciousness   32

II. THE DEPLOYMENT OF THE BIOSPHERE, AND THE SEGREGATION OF THE ANTHROPOIDS   37

    The launching platform of life: mono- or polyphyletism?   38

    1. Original characteristics of the biosphere   40

    2. The tree of life: its general shape   41

    3. The tree of life: search for the leading shoot: complexification and cerebralisation   47

    4. The Pliocene "anthropoid patch" on the biosphere   57

CONTENTS

III. THE APPEARANCE OF MAN, OR THE THRESHOLD
OF REFLECTION   *page* 61

    Introduction: the diptych   61
    1. Hominisation: a mutation, in the external
       characteristics of its appearance, similar to all
       the others   64
    2. Hominisation, a mutation that, in its develop-
       ment, differs from all the others   72

IV. THE FORMATION OF THE NOOSPHERE   79

    1. *The socialisation of expansion: civilisation and
       individuation.*
       Introduction: preliminary remarks on the
       notion of noosphere and planetisation   79
       1. The population of the World   82
       2. Civilisation   85
       3. Individuation   93

V. THE FORMATION OF THE NOOSPHERE   96

    2. *The socialisation of compression: totalisation and
       personalisation: future tendencies*
       1. An accomplished fact: the incoercible
       totalisation of man and its mechanism   96
       2. The only coherent explanation of the
       phenomenon: a convergent world   100
       3. Effects of, and forms assumed by, con-
       vergence   104
       4. The upper limits of socialisation: how
       to picture to ourselves the end of a
       world   112
       5. Final reflections on the human adven-
       ture: conditions and chances of success   117

Index   123

NOTE

As the very title shows, what follows makes no claim at all to give an exhaustive definition of man. It is simply an attempt to specify his "phenomenal" features, in so far as *the human* may legitimately (as observed by us earth-dwellers) be regarded by Science as the extension and crowning point of *the living*—provisionally, at any rate.

The aim of the present work is strictly limited: it is to try to define experientially this mysterious *human* by determining, structurally and historically, its present position in relation to the other forms assumed around us, in the course of the ages, by the stuff of the cosmos.

In distance and scope our objective is restricted; but what is absorbingly interesting about it is that it allows us, if I am not mistaken, to reach a privileged position from which we see, with deep feeling, that if Man is no longer (as one could formerly conceive him) the immobile centre of an already completed world—on the other hand from now on he tends, in our experience, to represent the very leading shoot of a universe that is in process, simultaneously, of material "complexification" and psychic interiorisation: both processes continually accelerating.

It is a vision whose impact should strike our minds with such force as to raise to a higher level, or even to revolutionise, our philosophy of existence.

Paris, 10th January, 1950

As its title indicates, this book sets out to study the structure of the human zoological group and the evolutionary directions it follows. This is simply another way of stating, and attempting once more to solve, the classic problem of "Man's Place in Nature."

Man's place in nature . . . we may well wonder why, as science progresses, this question becomes continually more important and fascinating to us.

In the first place, no doubt, it is for the eternal, entirely subjective (and therefore somewhat suspect) reason that the problem touches what is very close to us—our own selves.

Even more, however (and here there is no suggestion of anthropomorphic weakness) it is because we are beginning to realise in our minds—and this as a direct function of the most recent advances in our knowledge—that man occupies a key position in the World, a position as the principal axis, a polar position. So true is this, that to have understood the universe it would be sufficient for us to understand man: just as, again, the Universe would remain outside our understanding if we were unable coherently to integrate with it man in his entirety, with no distortion—the whole of man, I say advisedly, and including not only his members, but his thought, too.

We must, indeed, be very much blinded by the closeness to us of the phenomenon of man (within which we are immersed) not to be more vividly conscious of how impressively unique—even by virtue of its phenomenal nature—is this event.

Man is, in appearance, a "species," no more than a twig, an offshoot from the branch of the primates—but one that we find to be endowed with absolutely prodigious biological properties. Something ordinary: and yet pushed to even more than the extraordinary . . . Because it has been able to bring about such far-reaching effects in invading and transforming all that surrounds it, "hominised matter" (the only direct object of scientific interest), must surely contain within itself some prodigious force; it must be life carried to its very limit; it must, that is to say, finally represent the stuff of the cosmos in the most complete, most fully realised form known to the whole of our experience. When we remember that throughout the whole of a first scientific age (practically the whole of the nineteenth century) man was able to examine other worlds without ever appearing to be amazed at his own self, must we not admit that here, if anywhere, we have a case of the danger of the trees hiding the wood—or the waves the majesty of the ocean?

When looked at from too close, at the spatial and temporal scale of our individual lives, mankind too often appears to us as a vast, incoherent, restless movement around the same spot.

In the five chapters that follow, I shall try to show how it is possible, if we look at things from a sufficiently elevated position, to see the confusions of detail in which we think we are lost, merge into one vast organic, guided, operation, in which each one of us finds a place: a place that is, admittedly "atomic," but at the same time unique and irreplaceable.

Man making history meaningful.

Man, the only absolute parameter of evolution.

Five chapters, I said: and so five stages are envisaged, five phases selected to cover and picture the great spectacle of Anthropogenesis.

1. The place and significance of Life in the Universe. A self-involuting World.

2. The Deployment of the Biosphere, and segregation of the Anthropoids.

3. The Appearance of Man; or the "threshold of Reflection."

4. The Formation of the Noosphere.

    a. The expansion phase: Civilisation and Individuation.

5. The Formation of the Noosphere.

    b. The compression phase: Totalisation and Personalisation.

We must now try to examine these five points more fully, taking each in turn.

CHAPTER I

THE PLACE AND SIGNIFICANCE OF LIFE IN THE UNIVERSE

A SELF-INVOLUTING WORLD

---

i. PHYSICS AND BIOLOGY: THE PROBLEM

Man is a part of life. He is even (and this, in fact, is the thesis maintained in the pages that follow) the most characteristic, most polar and most living part of life. Thus, we cannot correctly appreciate his position in the world without first determining the place of life in the universe; this means that the first thing we have to do is to decide what life represents in the general structure of the cosmos: in so doing, we may still find it well, too, more or less deliberately to use such evidence as is provided by an examination of man himself.

In this first chapter, then, we shall (and indeed must) try to make up our minds about the meaning and value in the evolution of the universe, of the phenomenon of life. We must try, if possible, to build a bridge, or at any rate the skeleton of a bridge, between physics and biology.

Bearing this in mind, we can then, I think, best approach the heart of the problem by taking a concrete instance, going back in our minds to the time, some sixty years ago, when the Curies announced the discovery of radium. At that moment (though we have perhaps forgotten it) physicists found themselves faced with a puzzling dilemma. How were they, in fact, to try to understand this new element? In the discovery of this strange substance was science confronted with a particularly aberrant form of matter, or, on the contrary, with a new state of matter? Was it an anomaly, or a paroxysm? Had they simply found one more rarity for the curious to add to their collections, or did it mean that a completely new physics would have to be constructed?

In the case of radium, these doubts were short-lived. But when we come to the similar, but much more important case of life, we find to our surprise that a similar hesitancy still persists. If we try to "psycho-analyse" modern science, we are forced to this conclusion: that life, in spite of the extraordinary properties that make it unique in our experience, and in fact *because* it is so infrequent in appearance and on so small a scale (so ridiculously localised, for no more than a moment of time, on a fragment of a star), is still in practice looked upon and treated by physicists—as radium was initially—as an exception to or an irregular departure from the great natural laws: an interesting irregularity, no doubt, on the terrestrial scale, but with no real importance for a full understanding of the basic structure of the universe. That life is an epiphenomenon of matter—just as thought is an epiphenomenon of life, is, still too often, what too many people, implicitly at least, hold to be true.

It seems to me essential then to protest without delay against this depreciatory attitude, by emphasising that (again, as with radium) there is another solution to the dilemma presented by the facts to the intelligence of interested inquirers: that life is not a peculiar anomaly, sporadically flowering on matter—but an exaggeration, through specially favourable circumstances, of a universal cosmic property—that life is not an epiphenomenon, but the very essence of phenomenon.

We must note carefully this initial position, since everything that will be said later depends entirely on the frankness and resolution with which we make up our minds to take, intellectually, the next forward stride. Our position may be described as follows.

Everything goes to show that modern physics would never have been born if—against all possibility—physicists had obstinately persisted in regarding radium as an anomaly. Similarly, I shall maintain, biology cannot develop and fit coherently into the universe of science unless we decide to recognise in life the expression of one of the most significant and fundamental movements in the world around us. And this, moreover (here we reach the heart of the problem) not simply as a result of some emotional or gratuitous choice, but for a whole group of solid reasons that come to light as soon as we begin to realise the intimate structural link that connects the "accident of life" to the vast universal phenomenon (so obvious and yet so little understood) of *complexification of matter*.

This is something that must be clearly appreciated if we are to get away to a good start in our study of man and hominisation. First, however, we must, if we are to see our way clearly, define more precisely the terms we are using. Life, as I shall continue to insist throughout what follows, appears experimentally to science as a *material effect of complexity*. What then, in this particular case, is the exact, technical, meaning of "complexity"?

II. THE BASIC PROPOSITION: DIFFERENT FORMS OF ARRANGEMENT OF MATTER: "TRUE" AND "FALSE" COMPLEXITY

First, in what follows, I shall not, of course, use complexity to mean *simple aggregation*, i.e. any assembly of non-arranged elements—such as a heap of sand, for example—or even such as the stars and planets (apart from a certain zonal grouping due to gravity, and however numerous may be the substances that make them up).

Nor shall I use complexity to connote simple, undefined, geometric *repetition* of units (however varied they may be, and however numerous their axes of arrangement), such as we find in the astonishing and universal phenomenon of crystallisation.

I shall strictly confine my use of the word to the meaning of *combination*, i.e. that particular higher form of grouping whose property it is to knit together upon themselves a certain fixed number—whether great or small matters little—of elements, with or without the secondary addition of aggregation or repetition—within a closed whole of determined radius: such as the atom, the molecule, the cell, the metazoon, etc.

A fixed number of elements, a closed whole: this twofold characteristic of complexity must be emphasised, for on it depends the whole course of the thesis developed here.

In the case of aggregation and crystallisation, the arrangement continually remains, by its nature, incomplete. At any time a new contribution of matter is possible from outside. In other words, in the star or the crystal, there is nothing in the way of an inherent unity, confined within itself. All there is is simply the emergence of an accidentally "rounded-off" system.

Combination, on the contrary, produces a type of group that is structurally completed around itself at each moment (even though, starting from a particular class,[1] as we shall see, it is indefinitely extensible from within): the corpuscle, a unit truly and doubly "natural" in the sense that while organically limited in its contours so far as its own existence is concerned it also, at certain higher levels of internal complexity, manifests strictly *autonomous* phenomena. We find complexity progressively giving rise to a certain "cen-

[1] The class of "living corpuscles."

tricity"—not of symmetry, but of action. To put it more briefly and exactly, we might call it "centro-complexity."

Let us try to see, then, how the still hardly coherent universe of the physicists and biologists appears if we re-arrange it from top to bottom in the light of this concept of centro-complexity. (See fig. 1, below.)

III. THE CURVE OF "CORPUSCULISATION": LIFE AND COMPLEXITY

Fig. 1, you will note, is a curve constructed on two axes.

One of these, oy, calls for little comment. Basically, in the form I have given it, it is borrowed from Julian Huxley; it does no more than express, in centimetres, the length (or diameter) of the principal key-objects identified so far by science in nature, from the smallest to the largest.[1]

The second axis, ox, is more unusual. It is an attempt to express and measure not the linear dimension of things but (in the sense defined above) their degree of complexity. I hasten to add that this representation is more conceptual than actual, since once we go beyond molecules it rapidly becomes impossible (at least for the moment) to calculate either the number of elements (simple or compound) that make up a being, or the number of links between the elements or groups of elements. However, as a very rough approximation, we have used as a "parameter of complexity," in the case of the smallest corpuscles, the number of atoms grouped in the corpuscle.[2] This should suffice, I believe, to give some

---

[1] It would perhaps have been more in line with the latest views to have taken, in fig. 1, 10^{-13} as the point of origin of oy—since there is a possibility, according to some physicists, that that length may one day be found to be a (minimum) absolute quantum of length in the universe.

[2] Until we reach the living corpuscle, we could also use the molecular weight for the same purpose. When we go further (i.e. beyond the proteins) that coefficient ceases to be measurable or to have any precise meaning.

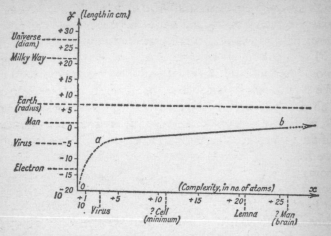

Fig. 1. Natural curve of complexities (see text).
a, point of Vitalisation
b, point of Hominisation

idea of the order of magnitude of the colossal numbers we shall have to accustom ourselves to deal with in this field.

Next, I have tried, using the two axes so chosen, to trace diagrammatically in its broadest trend, what I shall call the universe's *curve of corpusculisation*: the curve produced by grouping the *natural* corpuscles we know according to their two coefficients of length and complexity. This curve starts from the simplest infinitesimal (nuclear elements) and climbs rapidly to living corpuscles. Beyond that it climbs more slowly (for the increase in size varies relatively slightly in comparison with the increase in arrangement). I have drawn it as asymptotic to the radius of the earth, to suggest that the highest and greatest complexity achieved (so far as we know) in the universe, is what I shall later call planetised humanity—the noosphere.

22

Bearing the significance of this curve in mind, let us now examine it more closely and try to understand it. If we know how to interpret it, this is what we shall find:

The first thing it brings out is how much our universe would be mutilated if we reduced it to the very great and the very small—to no more, that is, than to Pascal's two "abysses." Even if we disregard the depths of time—if, that is, we simply take a section through the universe at an instant of time—there is still a third infinite: that of complexity. If we look at the figures on *ox*, we see that they reach astronomical proportions. . . . Spatially, then, the world is not constructed on two infinites (as is so often said) but on three at least; two of these, the infinitesimal and the immense, are generally accepted. But there is also (with its roots, like the immense, in the infinitesimal, but later branching off in its own direction) the immensely complicated.

This leads directly to a further and even more important point than that we first emphasised. Every infinite, physics teaches us, is characterised by certain "special" effects proper to that infinite: not in the sense that it is the only thing to possess them, but in the sense that those effects become perceptible, or even dominant, at the particular scale of that infinite. Such, for example, in the infinitesimal are the quanta, and relativity in the immense. Once we admit this, we have to ask what can be the specific effect proper to the vast complexes we have just recognised as constituting a third infinite in the universe. If we consider that question carefully, surely we must answer that the specific effect is in fact precisely what we call life—life, with its two series of unique properties: one, a series of external properties (assimilation, reproduction etc.) the other internal (interiorisation, psychism).

If my conclusion is correct, we reach here, in fact, the emancipating prospect on which depends for us the meaning and future of the world. The living, as I said above, has long been regarded as an accidental peculiarity of terrestrial matter. As a result, the whole of biology has been left out on its own, with no intelligible connection with the rest of physics. This is corrected if (as suggested by my curve of corpusculisation) life is, in scientific experience, no other than a specific effect (*the* specific effect) of complexified matter: a property in itself co-extensive with the whole stuff of the cosmos, but perceptible to us only where (after stepping over a number of thresholds that we shall later specify more exactly) complexity exceeds a certain critical value—below that value we cannot perceive it at all. The speed of a body must approach that of light for the variation in its mass to be apparent to us. Its temperature must reach 500° c. for its radiation to be visible to us. Is it not, then, reasonable to expect that through just the same mechanism matter, until it begins to approach a complexity of a million or half a million, should appear "dead" (though "pre-living" would be the better term), while beyond that figure it begins to show the red glow of life?

Once we take up this point of view—which makes of biology simply the physics of the very highly complex—it is interesting to see how everything included in our experience falls into place: and I mean everything, starting with the distribution and apportionment of the beings around us. If we look once more at our curve of corpusculisation, it is remarkable to find how readily it gives us the most flexible and most *natural* classification possible for the multiple units that make up the world we live in. If we follow *oy*, proceeding, that is, according to order of magnitude, the categories of objects succeed one another, and are mixed up with one another, in a completely incoherent way: there is no clarity. On the other hand, if we go by order of complexity, everything in this labyrinthine collection falls smoothly and effortlessly into place. Only the stars, as being simple aggregates, fail to fit into the scheme; and even in their case we cannot be certain that we may not soon be able to find some exact functional relation between moleculisation (corpusculisation) and astral condensation that will allow us to include them. For are not the stars and planets the crucibles in which, either by integration (of the more simple into the more complex) or by disintegration (of the more complex into the more simple) the various particles that make up our universe are produced? Without the earth could there be man?

Our curve, I repeat, gives us a *natural classification*. We are justified, therefore, in adding (and here we have the support of one of the widest and most definite conclusions of the very latest biological experience) that it gives us an *order of birth*, and so a *line of genesis*. In so far as it follows the pattern of the real, the curve in fig. 1 has a two-fold advantage. It groups the corpuscular types we see in our world to-day in a way that our minds recognise as coherent and logical; but it also brings out how those types have been formed successively throughout cosmic duration—as, indeed, the whole of modern systematic biology witnesses.[1]

The next thing we have to do is to form a more exact picture first of the general mechanism of this genesis (or rather cosmogenesis) and then of its hidden dynamism.

---

[1] Making allowance, where necessary, for momentary effects of dispersal or fanning-out, giving rise (as in a rainbow) to successions of types or objects that do indeed, when juxtaposed, produce a "natural series"— but which do not thereby represent the lines of progress, or trajectories, of successive stages gone through in the course of time: *spectra*, not descentlines of forms.

IV. THE MECHANISM OF CORPUSCULISATION: THE TRANSIT TO LIFE

Still looking at our curve in fig. 1, we find, as noted there, two main points:

a. The appearance of life, properly so called—by which I mean "formal perceptible life" (the point of vitalisation—or again, as we shall call it, the point of phyletisation).

b. The appearance of man (the point of hominisation or of reflection).

In this first chapter we may confine our attention to the pre-living segment of the curve, *oa*. But before we begin, I have an apology and an explanation to add. For some pages I shall have to enter a scientific field (physics and chemistry) that is not my own. I trust the reader will see in this intrusion not a claim on my part to solve problems that are outside my competence but simply a sort of appeal addressed by a biologist to his colleagues in the field of physics and chemistry. I am asking them, when they exercise their skill in analysis, to allow a progressively greater part to the evolutive or genetic point of view, for it is this that is most likely to reconcile their work with closely parallel work now being carried out in the domain of life.

After that digression, let us go back to the segment *oa* in my corpusculisation curve. On the diagram it appears extremely short. If, however, we consider the volume of matter involved and the duration of time occupied by this first emergence of cosmic complexity, we shall realise that it is in fact something quite colossal, even all-embracing, since, from the first origins of the universe, it covers the complete totality of astral matter. First, we have the whole transformation of atoms, and then that of molecules. . . .

1. *First, the formation of atoms*

One of the most curious intellectual phenomena to be produced in the field of scientific thought during the last fifty years, is without doubt the gradual, irresistible invasion of physics and chemistry by history: with the prime elements of matter exchanging their quasi-absolute mathematical position for that of contingent, concrete, reality; and with physics and chemistry, formerly branches of calculation, beginning to appear as preliminary chapters to a "natural history of the world": a strange reversal, indeed, of our picture of the universe.

No one doubts any longer that there has been a genesis of atoms, and that it is still going on. Astronomers and physicists, however, still seem far from unanimous about what type of genesis (simple or multiform) it may be. How are nuclei and electrons (themselves elements for which we shall one day have to discover or hypothesise the act that brought them into being)—how are they grouped, from hydrogen to uranium, in the various pigeon-holes represented by atomic numbers and their isotopes? Do they fall *directly* (as an effect of particular temperatures or pressures) into one or another pigeon-hole (a "spectrum-series") or should we, rather, conceive them (an "additive series") as assembling gradually, in stages, starting with hydrogen? Or, conversely, (a "subtractive series") as resulting, again in successive steps, from disintegration of matter initially ultra-condensed? . . . If I understand the position correctly, we know more, at the present moment, about how atoms disintegrate than about how they integrate.

One thing, at any rate, emerges from all this uncertainty: the only thing, in fact, that really concerns my present subject. It is this: that whatever conditions (as yet insufficiently determined) may later be discovered as governing the formation of atoms, in every case that formation shows, in comparison with things that have life, a differential characteristic on which our attention must be concentrated. By this I mean the *absence* of true lines of descent (or phyla). Whether atoms are formed in one single process or in several stages, throughout their history they undergo no more than—to put it at the highest—"ontogeneses." Each atom is ultimately born, whether slowly or more rapidly, only for itself—without transmitting anything: just as when a house is built. And the types of possible houses correspond to a limited number of predictable mathematical combinations. In spite, too, of the astonishing success of nuclear physics in the study of the transuranic elements, the atomic synthesis of matter seems to have reached a ceiling above which it can now make little advance. In this particular direction, the progress of corpusculisation seems to all intents and purposes to have been halted: though this does not prevent it from making an even more vigorous leap forward in another direction that offers a greater variety of choice: towards, in fact, the molecules.

2. *The genesis of molecules and living proteins*

From the evolutive standpoint we are adopting, one of the most curious characteristics of molecules is the way in which they show themselves capable of appearing, of "germinating," anywhere without exception in the world of atoms. There is no atom that cannot, under certain conditions, enter into molecular combination. Thus the molecular world does not *branch* off from the atomic world: rather, it envelops it, as would a cloud or an atmosphere—though this does not mean —far from it—that in certain sectors and along certain radial lines we do not find a particularly active and *additive* moleculisation: as happens, most noticeably, at low temperatures, starting with carbon. The world of atoms behaves as a sort of rigid assembly; the world of molecules, on the contrary, manifests a real internal plasticity, that enables it to "flow" freely and to push out sorts of "pseudopodia" in any favourable direction. Such, for example, is the remarkable group, on which we must now concentrate our attention, of the mysterious proteins.

By proteins, using the word in a very wide sense, I mean here the proliferation of substances so patiently and earnestly studied by organic chemistry, in which binary groupings, such as CO, CH, NH, are associated with various radicles, in a chain-association that may be simple or multiple, elongated or clustered, until fantastic molecular weights, up to several millions, are attained. This gives them an extraordinary flexibility of form—hence the pun of the "Protean proteins."

A serious difficulty that we meet when we study the "natural history" of the proteins arises from the fact that in the world of to-day we have no knowledge, or little knowledge, of them in a free state—but only in association with living beings; and we may suspect that it is because they are sheltered by these living beings and exist as a function of them, that, in the course of time, they have reached such a high level of super-complication.

This is a most awkward gap in our knowledge. That such a gap should exist at this level is only one more example of the strictness with which—as we shall continually have occasion again to note—direct perception of the origins of anything is automatically denied to our eyes as soon as a sufficient depth of the past is interposed. Even so, and in spite of this gap, it is impossible, given the present distribution of carbon compounds on the surface of the globe, not to assume that substances of the protein type must have been produced on the superficial, sensitive, irradiated zone of the infant earth. Following on from that, we cannot but guess that it was within those primordial proteins that—however staggeringly impossible it may seem, and yet by an almost inevitable effect of planetary geochemistry[1]—the great phenomenon of vitalisation must have been produced.

We shall see later that it was as a result of, and within, a coming together of primates that man must have been produced in the Pliocene. Similarly, it was thanks to and in the midst of a proliferation—we might almost call it a glow—of proteins that life on earth must have emerged, and burst into flame for the first time.

This very conclusion presents us with a final question.

In the case of man, we shall note later, it is with a revolution psychic in order (the appearance of the ability to reflect) that we may link the whole cluster of neo-properties that determine the formation of the noosphere. But now, in the case of the dawn of life, where are we to look for the fundamental mutation that we must suppose to have been produced somewhere, at some time, within the mass of carbonaceous terrestrial molecules, if some particular proteins, rather than others, were to be given the extraordinary opportunity of setting in motion the biosphere? Where, indeed, if not perhaps in the two-fold discovery of molecular dissymmetry and the mechanism of cellular assimilation?

This is an important point and merits close attention.

We saw earlier that the essence of true corpuscular complexity is that it appears in self-enclosed unitary groupings (unlike what happens, for example, in the crystal). Now, there are two different ways in which we may conceive such closed systems—according to whether they are found definitely arrested at their own level (as with a molecule of water or benzine) or, on the contrary, show themselves capable of modifying their composition, i.e. their complexity, without "unravelling" themselves (which is precisely the case with the cell). In this second type of corpuscle the unit remains,

[1] See A. Dauvillier, "Le Cours de physique cosmique du Collège de France," *Revue scientifique*, May 1945, p. 220.

indeed, at every moment closed in on itself, but it is with a mobile enclosure—for the complexity, too, can continue to increase at each moment without the particle being broken up.

In spite of their extraordinary elasticity, alternating (as isomers) between the crystalline and organic states, the "dead" (which really means "pre-living") proteins still belong to the first category, that of arrested corpuscles. On the other hand, however close the most elementary living things (viruses, bacteria) may still be to the proteins, what determines them is surely that they have contrived to leave a way continually open to a further increase of complexity and unified heterogeneity.

This is a very simple concept, but the more we think about it, the more, in fact, are we led to see the world of life as a vast sheaf of particles rushing headlong (through the operation of assimilation and its allied processes, association, reproduction, multiplication) down the slope of an indefinite corpusculisation: indefinite, and yet at the same time we are already perhaps beginning to see, ahead of us, the outline of its terrestrial term (see chapter v on the convergence of the noosphere). Earlier, we defined *a* in our curve as the point of vitalisation. We could equally well call it the point of phyletisation. Beyond that point we do in fact still find corpuscles becoming more and more rapidly and astronomically complicated. But, unlike what happened before, these corpuscles are constructed and subsist only *in a series* additively, each backed by the others—as though in a sequence or a continuous trajectory—each overlapping its predecessor—progressing towards a fulfilment still to be attained. So we find the whole of physics and chemistry recast and transformed by the discovery and development of phylogenesis.

There can be no doubt that this is what occurs. But for such a mechanism—we might call it "unfettered moleculisation"—to be initiated and to continue to operate, we must assume the existence and influence, underlying it, of some powerful dynamism.

In concluding this chapter, that is the point to be stressed.

V. THE DYNAMISM OF CORPUSCULISATION:
THE EXPANSION OF CONSCIOUSNESS

Our minds are now beginning to escape from the limitations of the static cosmos of antiquity, and to become familiar with the idea of major currents that affect the universe in its totality. First, there are the regressive currents: entropy, dissipation of energy—these were the first to be identified. But there are progressive, or constructive, currents, too. Do not scientists now speak of a universe that has been in process of explosive expansion ever since some primitive "atom" in which time and space were compressed in some sort of absolute zero?

If we wish to understand man, it is on this scale and along these lines, if I am not mistaken, that we should think of life.

We may put it this way: if we can explain the shift to red in the spectrum of the galaxies only by assuming a universe expanding in space, such an explanation is perfectly acceptable and no one has any objection to raise. When, therefore, we have to find an intelligible explanation of the persistent, insistent, ubiquitous mechanism of corpusculisation, why not assume a universe that, in one complete all-embracing whole, folds in upon itself until it is interiorised in a growing complexity?

I know very well, and feel it myself, that we are deeply influenced by the fact that from the old determinist point of view there is something improbable in the formation of the higher living complexes. In consequence, we feel an instinctive repugnance against forcing them all together into a scientific scheme of definite "causality." When we try to construct a physics of the organised, this idea of the exceptional and the abnormal continually appears. And yet the facts themselves—a continually growing accumulation of facts—must surely make us admit that:

"Without any doubt, one portion of the cosmic stuff not only does not disintegrate but even begins—by producing a sort of bloom upon itself—to vitalise. So true is this, that besides entropy (by which energy is dissipated), besides expansion (by which the layers of the universe unfold and granulate), besides electrical and gravitational forces of attraction (by which sidereal dust conglomerates), we are now forced (if we really wish to cover the whole of experience and include the *whole* phenomenon) to envisage and admit a constant perennial current of "interiorising complexification" that animates the whole mass of things".[1]

There we have a first point settled. Quite apart from any scientific (still less finalist) explanation we may offer, the universe, as though "ballasted" with complexity, falls from above into continually more advanced forms of arrangement.[2]

[1] In this connection, we could say that the two axes in fig. 1, *oy* and *ox* (taken not as axes of co-ordinates but as axes of movement) correspond to the two main directions of cosmic evolution: on one hand, along *oy*, the universe expanding from the infinitesimal to the immense; on the other, along *ox*, the same universe folding in upon itself and centring on itself, from the extremely simple to the immensely complex. In both cases the movement does not slow down but (as though continually falling forward) accelerates.

[2] The cosmic slide from the simple to the complex (or, which comes to the same thing, from the unarranged to the arranged) corresponds, we may note, to the passage from an unordered to an ordered heterogeneity—and not, it should be emphasised, to a Spencerian passage from the homogeneous to the heterogeneous. The initial multiple can be conceived only as an immense scattered diversity. Here we may note incidentally that there may be a hidden relation between the Newtonian gravity of condensation (which produces the stars) and the "gravity" of complexification (which produces life). In any case, the two can function only conjointly.

To do no more, however, than bluntly state the fact will not satisfy our minds, insatiably eager to get to the bottom of the problem. So far as actually existing is concerned, a cosmic movement of folding in upon itself seems quite undeniable. Where, however, are we to place its driving force?

Here we have three possible intellectual points of view.

a. Should we first (this is the *materialist* line) ascribe the enigmatic power of corpusculation to an automatic force of natural selection, *sui generis*, which drives matter (when it has succeeded, by the statistical operation of chance, in escaping from disorder and simple crystallisation) first to plunge over, and then, snow-balling, to roll with increasing momentum down the slope of a continually increasing complexification?

b. Should we, on the other hand (this is the *spiritual* line) look for it in an "expansion of consciousness"—consciousness[1] striving irresistibly (like an idea in the mind) to attain its maximum fulfilment, but unable to do so unless it can continually and increasingly, by process of invention, arrange, i.e. centre, matter around itself? This means not, as in the first explanation, "a continual increase of consciousness in the world, because of a continual increase in complexity" (achieved by chance), but "a continual increase of (planned) complexity, because of a continual increase of (gradually emergent) consciousness."

c. Finally, should we (dissociating ourselves from the spirit-matter conflict) do no more than put it as follows? In the older universe of Laplace, the quantity of contingence, once initially posited, remains indefinitely the same in any subsequent state of the system whatever may be its indefinite transformations. In the universe of Einstein, on the contrary, or Heisenberg, the quantity of indetermination (because it is

[1] Consciousness, i.e. *the within* (sometimes experimentally apprehensible, sometimes, because it is infinitesimal, inapprehensible) of both pre-living and living corpuscles.

continually fed by the action of each corpuscle) varies, and a better arrangement of the system may cause it to increase. In that case, wherever vitalisation of matter is possible would it not provide some sort of overflow for this continually increasing mass of indeterminate secreted by the universe?

It will, I hope, become clear from what we say later (cf. chap. V, p. 109) that if, until we come close to man, the determinist driving force of natural selection may, at a pinch, be sufficient to account externally for the progress of life—yet at least from the "threshold of reflection"—certainly not later—we must add to it, or substitute for it, the psychic power of invention: only thus can we explain the ascending progress of cosmic corpusculation right up to its higher termini.

On this point, no doubt, science has not yet said its last word.

In every case, however, this at least remains true (and this, basically, is the only question that matters here) that if our world is indeed a thing that is characterised by arrangement, in one way or another, then we can better appreciate that life can no longer be regarded as a superficial accident in the universe: we must look on it as (under pressure everywhere in the universe), ready to seep through the narrowest fissure at any point whatsoever in the cosmos—and, once it has appeared, obliged to use every opportunity and every means to reach the furthest extremity of everything it can attain: the ultimate, externally, of complexity, internally of consciousness.

It is this that makes the study of man and his genesis, on which we must now embark, so fundamental and so dramatic.

Man: not simply a zoological type like the others. But man, the nucleus of a movement of in-folding and convergence in which, localised on our little planet (lost though it be in time and space) is manifest what is probably the most characteristic and most illuminating current that affects the immensities that envelop us:

Man, on whom and in whom the universe enfolds itself.

CHAPTER II

THE DEPLOYMENT OF THE BIOSPHERE AND THE SEGREGATION OF THE ANTHROPOIDS

In the last chapter we were studying what I have called the "curve of cosmic corpusculisation," and stopped at point *a* of vitalisation (or phyletisation). It was here, I said, that starting with certain proteins endowed with the mysterious power of "assimilation," matter was caught up in a process of super-moleculisation constantly opening out ahead. In this second chapter we have to extend our analysis to the segment *ab* (see fig. 1), but excluding point *b* itself (the point of hominisation or reflection), which we shall study in a later chapter. Even with this limitation, it is a vast subject, disproportionately vast, one might almost say, since this "small" segment represents in reality the incredibly complicated fascicle of genetic fibres (phyla) developed over a period of six hundred million years. . . . Nevertheless, if only because it is so vast, it is a subject that we will do well to try to embrace in one sweep, reducing it to its most important structural elements.

To this end, I propose, after some remarks on what one might call the presumable dimensions and explosive character of point *a*, the vitalisation point, to take the following points in turn:

1. The probable original aspect of the biosphere.
2. The tree of life: its general shape.
3. The tree of life: where do we find its leading shoot (complexification and cerebralisation)?

4. The axis of the primates and the "anthropoid patch" in the Pliocene.

So, to begin with our preliminary remarks.

THE LAUNCHING PLATFORM OF LIFE: MONO- OR POLY-PHYLETISM?

In fig. 1, the starting point of life is diagrammatically represented by a critical point, but this is only symbolic. What surface extension or even what structure are we to attribute to this point in the physical reality of things? By this I mean that if we want to see the facts as they occurred, in what number and following what rhythm are we to suppose the molecules of protein underwent the particular mutation that vitalised them? Was it in single units or in tens of thousands? And if it was not confined to a single explosive point, then in how many places and at how many different moments did it occur? In other words, when we look at life in its very first origins, should we consider it is monophyletic or polyphyletic?

We must recognise before we go any further that this is a question we cannot yet answer with any certainty and no doubt never will be able to. As I shall soon have occasion to emphasise, when we come to deal with the first appearance of man on the earth, the "beginnings," in every field, are lost to us: the past swallows them up and our eyes can no longer decipher them. Even in man's brief history this law operates strictly, and one can well imagine how it must apply in the case of an event so profound and affecting such infinitesimal elements as the animation of the first carbonaceous molecules.

One curious fact, however, has been noted that may allay the disquiet of our imaginations and help to restrict the dimensions of the problem: by this I mean the singular similarity to be observed between living substances on points so special and accidental that their resemblance in this case seems much less the result of some convergence than evidence of a real relationship. For example, in living beings, molecular dissymmetry is regularly found in only one of the two forms that the chemical elements might equally well have adopted. In protoplasm, glucose, cellulose and amino-acids all are dextro-rotatory; the albumins, cholesterol and fructose, are laevo-rotatory. Similarly, enzymes are found to be same throughout the whole series of living beings. How are we to explain this coincidence, or "unity of plan," in detailed characteristics? Should we see in it, as in the "pentadactyl tetrapody" of land vertebrates, an indication that at its very beginnings life germinated on a peduncle that was relatively narrow in section, in a more or less limited area of the earth, and by one single emission in duration? Or can these crystallo-chemical similarities be reconciled, on the contrary, with a wide initial starting area and the repeated influence of instances of selection and convergence?

I shall not attempt to answer that question, which, in any case can safely be left open. At this point in our inquiry only one thing really matters: and that is to realise that in either case (i.e. whether initially there was only a single point of vitalisation or *n* points) the result must have been the same. There must, I mean, have been an astonishingly rapid invasion of the whole photochemically active surface of our planet. It is as though the surface had then been, in relation to life, in a state of almost super-saturation, in consequence of which those elements in it that were capable of being vitalised were rapidly enclosed in one single membrane— the first elementary form of what in the course of geological periods was to produce the "biosphere."

I. ORIGINAL CHARACTERISTICS OF THE BIOSPHERE

By biosphere we mean here not, as some mistakenly do, the peripheral surface of the globe to which life is confined, but the actual skin of organic substance which we see to-day enveloping the earth: in spite of its thinness, a truly structural layer of the planet, a sensitive film on the heavenly body that bears us—and an admirably adjusted device in which, if we know how to look at it, we may see the bond (as yet rather felt than fully understood by our minds) that holds together biology, physics and astronomy within the same cosmic dynamism.

It is probable that in the very beginnings, which is where we are imagining ourselves, the biosphere did not extend beyond the liquid layer of the primordial ocean: though we cannot, indeed, be sure whether, in those far distant ages, even the smallest trace of some proto-continent was yet emerging from the waters.

What we do know is that, from the very beginning, the protoplasmic scum that appeared on the surface of the globe must have displayed, besides its "planetarity", the other characteristic that was to become regularly more pronounced in it in the course of ages: the extremely close interconnection, I mean, of the elements that made up this still shapeless, floating mass. For complexity cannot develop within each corpuscle without entailing a parallel and progressive network of relations, a delicate and ever-shifting balance, between neighbouring corpuscles. This collective inter-complexity is a natural extension and augmentation of the intra-complexity proper to each particle; and we shall later have to consider, in man, where it appears in the form of "social convergence," a remarkable manifestation of it, terminal and unique. For the moment we may simply note that however granular and discontinuous the layer of vitalised matter may have been initially, a network of affinities and deep-seated attractive forces (destined to become continually more pronounced) was already bringing together—and seeking ever more closely to compress upon itself in one vast symbiosis, this unnumbered horde of particles so charged with germinal power. Not indeed that it was a mere horde or agglomeration—for already, under the slow continual pressure exercised by the closed curvature of the earth, it was a close-woven web—within which the manifold arborescences were covertly beginning to appear. It is the characteristics of these that we must now try to distinguish, before going on to discover whether their apparent disorder does not conceal, besides a general polarisation towards greater complexity and consciousness, some principal axis of growth and consciousness.

2. THE TREE OF LIFE: ITS GENERAL SHAPE

In fig. 2, below, I have tried to express symbolically (but in an extremely simplified form) the main structural lines of the biosphere, as distinguished, in two centuries of work, by the patient and minute dissection of a whole army of zoologists and botanists. It is, remember, a simplified diagram, "projected" or "developed" on an imaginary flat surface, since, in the reality of nature, the ramifications shown have continually at every moment formed, both biologically and spatially, a whole closely rolled up, or you might say "clustered," on itself. We should note a further point: the original and immediate purpose of this diagram when drawn up by taxonomists was to cover only those species that make up the biosphere at the present time. In this case, however, as with fig. 1, it happens (as palaeontology confirms) that the morphological arrangement of types corresponds exactly to their chronological appearance in the world.

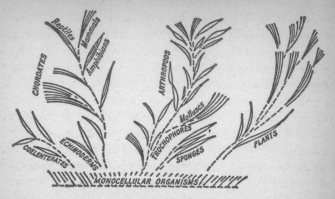

Fig. 2. The Tree (or arborescences) of Life. Simplified diagram (see text)

From this it follows that the tree of life, as represented here, can be regarded equally well (as happens with every *natural* classification) as expressing either the diversity of forms living in the present, or the history of their appearance in the past— the latter aspect, of course, being the one we shall be particularly concerned with.

With those explanations in mind, we may proceed without delay to a successive examination of the various elements in the diagram. At first sight there appear to be two sharply defined zones contrasting with one another: below, there is a confused matted, fibrous mass of monocellular beings, and above a highly ramified system of multicellular organisms.

A. *The monocellulars*

Still from the evolutionary point of view we shall adhere to throughout this book, the world of the unicellulars has this fascination about it, that it discloses and expresses, almost tangibly, the corpuscular origins and nature of life. Whether indeed we stop to note the simplicity of the smallest organisms as yet distinguished by the microscope (not more than a hundred molecules of protein in a bacterium one thousandth of a millimetre in length, and only a single one, may be, in the ultra-viruses and genes . . .), or whether we try to appreciate the fantastic proliferation of protozoa and protophyta that fill the fresh and salt waters of the earth, the pseudo-barrier that perhaps in our minds divides the unity of a mammal and that of an atom into two irreducible categories, tends equally to disappear. On the strictest experimental showing, when life emerges from matter, it is still redolent of a molecular state that it cannot but foster by the amazing activity of its power of multiplication.

While realising that, we should hasten to add that in spite of a very real "primitiveness" that no one denies them, the present monocellulars (as, in ethnology, is true of modern pre-civilised men) give us only a very imperfect idea of what their "fauna" could have been like in the first periods of their appearance. In their present form, we see them associated in a highly differentiated group of great antiquity, in which ultra-complex types (ciliates and shelled) are found side by side with other ultra-simple forms (viruses) in which we may be justified in seeing no more than degraded types. Moreover, at a time, too, that was probably very close to their origins, an important cleavage must have taken place in their initially homogeneous though confused mass: this separated the proto-plants (feeding on chlorophyll) from the proto-animals (parasites on the former)—not to mention the more mysterious group (whose development was halted) of autotrophic beings, those that are capable of assimilating "mineral" directly without the intervention of solar radiation.

It is from this initial cleavage that we can now climb a further step, into the world of the multicellulars, both vegetable and animal.

B. *The multicellulars*

Reduced to its essentials, and detached from the vast trunk of the vegetable kingdom round which it is twined (and with which we need not concern ourselves here) the animal world of the metazoa displays to-day two particularly active main stems, each of which (as has often been pointed out) represents one of the two major solutions to the problem of life.

On the one hand, that of the arthropods (Arachnida, Crustacea, insects) with an external carapace or skeleton: on the other, that of the chordates or vertebrates, with a mainly internal skeleton: these latter emerging at some time from their fish-like swimming forms, to produce the exceptionally "monostructural" group, progressive and mastering, of the walking tetrapods. In this group, which won undisputed dominance of the continents, we have confined ourselves (in the diagram) to distinguishing only the three major sub-groups, grafted on one another, the amphibians, the reptiles, and the mammals.

Outside and "below" these dominant stems, and with no well defined relationship with them, other sub-worlds again, extremely extensive but much less progressive, may be seen streaming out. On one side we have the trochophores (annelids, molluscs), more akin to the arthropods; on the other, the still more divergent echinoderms, coelenterates, sponges: a sort of background to the picture, or undergrowth, that gives evidence of the astonishing "creative" fertility and incredible power of proliferation with which the infant biosphere was endowed.

Let us, then, leave our brief inventory of the major zoo-logical types at this point, and now try to obtain a general view of the situation. Simply from the angle of "positional zoology" the diagram we have been looking at tells us, in short, three main things:

44

of cephalisation. There can be no doubt that we are on the right road and have only to keep straight on.

b. Let us, then, take a further step. Let us, that is, without leaving the mammals (but this time with the assistance of palaeontology) try to discover whether the progress in cerebralisation, characteristic of vertebrates in general, may not be occurring, recognisably and with a measurable gradient, within the group itself, even in the detail of a single phylum. Such a study has recently been carried out by an American palaeontologist, Tilly Edinger, for the family of Equidae. Everyone has heard of the classic line of descent of the horses, studied and re-studied on countless occasions but always, hitherto, in connection with the development of hooves, teeth and muzzle. Using this exceptionally well-marked phylum[1] Miss Edinger had the happy inspiration of investigating, by means of a large number of endocranial casts, how in the course of time its brain could have evolved from one age to another. This is indeed an impressive study, since in this case it is a question of following and analysing a movement that covers fifty-five million years. . . . The principal results are shown in fig. 4 below: and from this, among other things, we learn three things in particular.

1. Taking the whole, as we ascend the phylum, there is a clear accentuation of cerebralisation, and that in the particular way we mentioned above: development of the hemispheres, with an accompanying reduction of the olfactory lobes or rhinencephalon; multiplication of folds, increasing the surface area of grey matter; and a tendency to cover the cerebellum.

2. Initially, the brain is still remarkably primitive: the hemispheres are little developed, and almost smooth, as in an insectivore.

[1] A complex phylum, of course, itself made up of numerous overlapping lineages (cf. above p. 45).

Fig. 4. Development of the brain in Equidae (after Edinger).
Approximate time-span, 55 million years
  1. *Eohippus*, lower Eocene; 2. *Mesohippus*, middle Oligocene;
  3. *Merychippus*, middle Miocene; 4. *Pliohippus*, middle Pliocene;
  5. *Equus*, Pliocene
Note the delay and slowness of cerebralisation in the initial stages
(the brain of *Eohippus* is still at the lowest marsupial stage) and the
rapid progress that starts with the Miocene

3. The start—a rapid, almost revolutionary start—of cerebralisation (from *Mesohippus* onwards) is distinctly out of
step with the evolution of limbs. In spite of the backwardness
of his brain, *Eohippus* is already (notwithstanding the number
of his digits) a true "little horse."[1]

Thus, if it is followed along one and the same strand (provided we do so for a sufficient number of millions of years)
cerebralisation—understood in the precise technical sense of
the "development" of a neocortex or neopallium—not only
persists among the higher vertebrates, but very markedly
accelerates. Broadly speaking, with the mammals, we are
in a particularly active zone of cosmic complexification or
corpusculisation: and that means, to go back to our metaphor,

---

[1] This fact suggests that the particular superiority to which the mammals
owe their initial triumph over the reptiles in the biosphere is to be found
not so much in some cerebral mutation (as in man, see below, p. 62) as
in a physiological modification affecting circulation and reproduction
(isothermy, viviparity).

we have in them a well-defined leading shoot to the tree of life.

May there not, however, be some way now of pin-pointing this leading-shoot more exactly, not simply in a sub-class, but in an order or even (why not, indeed?) a single family?

And this is where the primates come on the scene.

C. *Second result obtained from applying the parameter of cerebralisation: it is through the order of primates, and more precisely through the family of anthropoids that runs the terrestrial axis of corpusculisation*

While the Equidae are primarily runners (as, among other animals, there are carnivores, swimmers, climbers or burrowers) the primates are in the first place "cerebral" creatures, or, if you prefer the term, "cerebromanual": the two going together. In their case (and in this it is unique) the particular orthogenesis of the phylum coincides with the general orthogenesis of life. It would therefore be supremely interesting to be able to reconstruct their brain-history in the same detail as with the horses. Unfortunately, for reasons familiar to palaeontologists, fossil remains (and specially crania) are particularly rare for this group of animals, except in the case, itself exceptional, of deposits in fissures and caves representing ancient dwelling-places.

In spite of these unfavourable conditions, there is a sufficient number of indications to prompt the belief that since the Eocene the cerebralisation of the primates pretty well, in its main lines, matches that of the Equidae. The endocranial cast of *Adapis*, in particular, with its "insectivorous" simplicity, corresponds remarkably to the *Eohippus* stage. In the same period, it is true, other forms are known (*Necrolemur*, tarsiers) whose globular head suggests the idea that from the Lower Eocene the primates, at least in some of their families, were more advanced in cerebralisation than the other mammals.[1] Whatever these precursors were like in reality, one thing is clear: that once the primates (like the Equidae, and at practically the same epoch) had entered the accelerated phase of their cephalisation, then (even if we exclude man) they travelled faster and further along that line than any other living creature around them. To realise how true this is, one has only to look at the hemispheres in the most primate of primates —by which I mean the anthropoids (or anthropomorphs)— and see how they are now, with their extensive folds and convolutions, coming to cover the cerebellum completely; and this characteristic, acquired apparently as early as the Miocene, is accompanied in them all by a remarkable over-all size of head: a size that certainly, even though no precise indication can yet be attached to it, must nevertheless have some significance.

In fact, once it is admitted that, in higher living beings, it is the degree of cerebralisation that measures *true* complexity (i.e. the absolute state of vitalisation), it becomes almost a truism to conclude that before man the principal axis of the cosmic movement of corpusculisation ran through the primates, and more particularly through the anthropoids. Here, as often happens, science does no more than elaborate and recast what has always been intuitively held by the ordinary layman.

With this conclusion to give us confidence, let us for a moment leave anatomy for geography. By this I mean that now that we have recognised, on precise morphological evidence, the biologically central position of the primates, we may try to follow, if only very summarily, the vicissitudes

[1] In the only endocranial cast of *Necrolemur* so far described (J. Hürzeler, "Zur Stammesgeschichte der Necrolemuriden," *Mém. suisses de Paléontologie* vol. 66, 1948, pp. 33*ff.*), the characteristics are somewhat contradictory: the hemispheres are relatively very large and rounded but completely smooth, not covering the rhinencephalon, which still distinctly projects in front of the brain.

of their expansion over the world, from the first time they enter our field of vision until the threshold of the point of hominisation.

The advantages of this shift of approach will soon be apparent.

4. THE PLIOCENE "ANTHROPOID PATCH" ON THE BIOSPHERE

Even though, as a result of the scarcity of fossil evidence, our osteological knowledge is still sadly deficient when it comes to the limbs and skull of the ancient primates, on the other hand we have a good many of their teeth and jawbones; these, again, are sufficiently characteristic for us to be able to use their evidence to recognise from era to era, starting with the beginnings of the Tertiary, the presence of the group in the different continents of the globe, and determine the general state of its development.

In its essential features, this bio-geographic history may be reduced to the five following phases:

a. *First appearance, in the Lower Eocene*, over a vast area including, simultaneously, North America and Western Europe: those two regions being at that time, it seems, connected by some North Atlantic bridge.[1] Extremely small forms (hardly larger than a mouse), some of them (the anapto-morphidea) decidedly "tarsioid." It would obviously be of the highest importance to know what was happening at the same time south of the Tethys.[2] Unfortunately we still have

---

[1] This is a more likely hypothesis than that of trans-asiatic communications, for the existence of which there is no positive palaeontological evidence.

[2] In Greek mythology Tethys was the sister and consort of Oceanus. The name was given by the Austrian geologist Eduard Suess (1831-1914) to the ocean that at one time stretched from Gibraltar to the East Indies. Tr. note.

not found any continental fossiliferous deposit of this era in Africa.

b. Increase in size and numbers, during the Middle Eocene. During this period, apparently, general conditions (both zoological and geographical) showed little change for the primates: the same types (lemuroid and tarsioid) spread over the same area. Nevertheless certain profound transformations were either in preparation or occurring. For one thing, the transatlantic bridge had already, it seems, been cut; for another, South America is being invaded—as established by the conditions met with at the beginning of the next phase.

c. Disjunction and radical transformation of the group during the Oligocence. Nothing further, definitely, in North America; and in Western Europe no more than the survival of some lemuroids. On the other hand, the establishment of a platyrrhine bloc in South America; and the emergence in Africa (Fayum) of an extremely lively evolutionary centre (an autochthonous focus, rather than one kindled by sparks come from Europe): appearance of the first anthropoids.

d. Expansion of the anthropoids in the Miocene. Starting from its African (and most probably Central African-Kenya) focus, the "anthropoidal" pulsation, headed by *Dryopithecus*, was at this time spreading widely over the whole southern edge of Eurasia. To the west, above the Tethys sea (by this time silted up) it reached Spain, France, and Southern Germany. To the east, although as yet we lack direct proofs of this, it probably spread as far as the Pacific, at the end of the Indian Ocean (though without, in the north, crossing the Himalayas and the Yangtze). After this, the western portion of the wave fell back to the south of what is now the Mediterranean, while at other points it consolidated and rooted its hold. This process ended in what one may call:

e. The establishment, in the Pliocene, of an anthropoid province. In nature as we know it now the large anthropoid apes (gorilla, chimpanzee, gibbon, orang-utan) constitute only a broken series of isolated groups from the Gaboon to Borneo. Since the end of the Tertiary, man has intervened here. On the other hand, judging from the frequency and distribution of the fossils we have, we must imagine a dense continuous layer of different types of anthropoids, covering, towards the beginning of the Pliocene, a wide tropical and sub-tropical belt running from the Atlantic to the Pacific. Teeth and jaw-bones of different anthropoids are relatively common in the sub-Himalayan deposits of that period; and we know that at the beginning of the Quaternary, there were still great numbers of orang-utan in Southern China and Indo-China.

Let us, then, stop for a moment and take a look at this so curiously inhabited area of the globe, and try to understand the extraordinary intensity that emerges from this particular time and place.

At first glance, one would say that the scene presents nothing of special interest: what, in fact, is there more to admire in this Pliocene triumph of the primates than in any other of the successful extensions of animal life won in various other places by this or that living form in the course of peopling the earth?

And yet, in the light of the principles that have guided us so far in our inquiry, from the corpuscular origin of the universe to this dawn of the modern world, can we not distinguish something profoundly symptomatic, and even dramatic, underlying the apparent ordinariness of the scene? Is it not apparent that the "area of anthropoid extension" is, by some chance, an area of maximum cerebralisation and hence of vital pressure? For a moment one might have thought the cosmic current of "complexification" lost in a confusion of layers, in the sands of the biosphere; and now we see it reappearing more clearly defined than ever, and canalised henceforth in a chain of neurones: it is now not only zoologically individualised in a particular family of primates, but, what is more, spatially localised—like the germ-spot in an egg—in a determined area of the world.[1] Throughout all geological ages an ever-increasing quantity of nervous substance has continually been isolating itself (and continually perfecting its arrangement) at the heart of vitalised matter. Now we see it, in its most highly developed form, coming together in a geographical association. This, surely, is an indication that some great event in planetary biochemistry is in preparation?

Earlier (chapter 1, p. 30), when we were trying to reconstruct the features of the infant earth, we found that we had to picture to ourselves certain assemblages or waves of proteins, floating on its surface, of which we could say that they were the "glow" of life: and now, six hundred million years later, quite close, in short, to our own time, the phenomenon is reproduced at a higher level. For anyone who has eyes to see it, the "anthropoid patch in the Pliocene" itself, too, "glows" under the influence of a new ascending radiation.

And it is, in fact, somewhere in this active continental zone that, in our next chapter, we shall see—across a major threshold of cosmic convolution and interiorisation—the emergence of thought, above, and as a new envelope to, the biosphere.

---

[1] An area sufficiently vast to allow an intense *simultaneous* multiplication both of the general population *and* of the isolated population groups of the primates in question: the first condition increasing, as an effect of the greater total volume, the chances for the appearance of the "hominising mutation," and the second, as an effect of segmentation, of its preservation.

CHAPTER III

THE APPEARANCE OF MAN,
OR THE THRESHOLD OF REFLECTION

INTRODUCTION: THE DIPTYCH

Among the innumerable contrasts presented to our minds as the panorama of geological ages unfolds, I know none as exciting, both because of its comparative closeness to us and because of its suddenness, as that which differentiates Pliocene earth from the modern earth. Try just to imagine, like two pictures set side by side, first, a sufficiently stable continental region (for example the Paris basin) a little before the Villafranchian, and then set against it the same area as we see it now. What does each picture show us?

In the former—towards the end, that is, of the Pliocene— the topographical and climatic setting is, in its main lines, the same as at present: the Seine, the Loire, the alluvial deposits radiating from the Massif Central, all under a temperate sky; and, apart from the large fauna that have disappeared (elephant, rhinoceros . . .) the animals (wolves, foxes, weasels, badgers, deer, wild bears) all belong to types still extant. It is, in fact, almost *our* world. And yet it is a world with an uncanny feeling that something enormous is missing. The setting is familiar, indeed, but there are no men—not a single man in sight. So total is this absence, that if by some miracle a traveller had been transported to our planet at that period (that even so is not so very distant—a couple of million years or so in the past), he could have covered the whole earth and met *nobody*:

I mean, *literally nobody*. We should try to appreciate the full sense of strangeness, of exile, of loneliness contained in those bald words.

And over against this (on the modern half of the same diptych) what do we see but men everywhere, a super-abundance of man, man cluttering the whole prospect with his houses, his domestic animals, his factories—man inundating like a flood the whole countryside and every remnant of wild fauna.

Faced with so radical a change brought about in so short a time, one cannot help asking what it was that happened between these two states of the earth, these two periods of time (that are yet geologically so close to one another), to make such a transformation possible: what catastrophe or what profound alteration in the governance of evolution.

We had a similar question to answer in similar circumstances (the emergence of the biosphere) at the very first beginnings of life. We had then to find a reason for the lightning expansion over the earth of the first membrane of organised matter, and our answer was that "there can be no doubt that certain proteins chanced to meet with the structure that allowed them to 'assimilate'."

In the present case, we shall have to link the "phenomenon of invasion" with a mutation of the psychic order, and state (for reasons that can be verified positively) that what explains the biological revolution caused by the appearance of man, is an explosion of consciousness; and what, in turn, explains this explosion of consciousness is simply the transit of a specially favoured ray of "corpusculisation" through the hitherto impenetrable surface that separates the zone of direct psychism from that of reflective psychism.[1] Once life, along

---

[1] Had some other zoological ray, by chance, crossed this critical surface before, there would never have been man: for it would have been that other ray that then blossomed into the noosphere.

this particular ray, reached a critical point of arrangement (or as we call it in this context, of convolution) it became hyper-centred upon itself, to the point of becoming capable of foresight and invention.[1] It became conscious "in the second degree." And this was sufficient to enable it in the course of a few hundreds of thousands of years to transform the surface and appearance of the earth.

In the two chapters that follow I shall concentrate on tracing, particularly in the field of socialisation, the progress of this psychic *reflection*, in which we see around us, in nature, the expression of the latest and no doubt supreme efforts of complexity.

As a start, however, we may confine ourselves in this chapter to a study of the observable conditions in which this tremendous transformation could, most probably, have been brought about—brought about, moreover, at a time so close to our own. In other words, where are we to position and how are we scientifically to characterise, *the threshold of reflection*?

It is a nice question, and complex; and it entails my elaborating a double series of mutually balanced considerations, that fall under two heads:

1. In the eyes of science, the appearance of man followed, essentially, the same mechanism (geographical and morphological) as every other species.

2. Nevertheless, right from his origins, we find in man certain special properties that denote in him a higher vitality than that we meet in other species.

---

[1] And, of course, of all that follows in consequence in respect of thought as the discoverer and builder of the world.

I. HOMINISATION: A MUTATION, IN THE EXTERNAL CHARACTERISTICS OF ITS APPEARANCE, SIMILAR TO ALL THE OTHERS

"Essentially, man appeared in just the same way as every other species." What exactly does that statement mean? There are, as we shall see, a number of positive meanings. But, to begin with, there is also a negative one; it may even be disappointing, but we must face it once and for all if we hope to be spared much wasted effort and fruitless dreams in our study of human palaeontology. It is this: just as with any other living form, we must realise that the very earliest human origins, from their very nature and however much we magnify the little we can get hold of, can never be the object of direct experiential knowledge.

I have already had occasion to mention in passing (cf. pp. 29, 38) the sort of fatality that seems, where we try to reconstruct the past, to take a malignant delight in obliterating just that particular point in things that we would most like to know—I mean their beginning. The origin of an intuition or an idea—of a language or a people—and *a fortiori* of a species or a zoological layer . . . you can never get hold of the real beginning of anything.

The more one thinks about this apparently fortuitous condition that governs our experience, the more one comes to realise that it represents in fact a profound law of "cosmic perspective" from which nothing can be immune: the selective result of absorption by time of the most fragile (the least extensive) portions of any development *whatsoever*. The embryos, whether of an individual or a group, of an idea or a civilisation, are never fossilised.

This being so, it is obvious that in the depths of time at which the zero of anthropogenesis lies (and we are concerned here with a distance geological in order) we must be prepared to meet a serious "blank" in our picture of the past. How, in fact, can we hope to find traces of the very first men when we have to admit our inability to know the first Greeks or the first Chinese? In such a case, the most that the laws of historical perspective allow us to hope is that we may be able to reduce, down to a certain minimum, the radius of the area of uncertainty (of indetermination) within which a point we cannot grasp lies hidden—the source of the stream we are trying to trace back to its origin.

However, even though it is of the very nature of the point of human emergence that we cannot grasp it in itself, in its concrete reality, nevertheless there is nothing to prevent us from determining indirectly its features (by which I mean certain of its properties and characteristics) by analysing the radiation that spreads out from it. We accept that in its exact geographical localisation and morphological forms, the hominising mutation will always elude us, but on the other hand the converging investigations of prehistorians are gradually revealing to us the infancy of man. This is enough to enable us to judge that, in its main lines, hominisation initially operated in accordance with the general law of all *speciation*, which is to bring about the emergence of living groups ramified in their over-all appearance, and in a state of active division.[1]

It is this precise point that I hope to bring out in the first part of this chapter, basing my attempt, initially, on what seems to me to be the true significance of *the "prehominians" of the Far East.*

[1] We hardly need to recall in this connection that since palaeontology can discern species only at the *group* stage, and that always at some considerable distance from their point of origin, the question of an original single *couple* (monogenism) has no scientific relevance. At so great a distance in the past, our scientific vision of life can distinguish nothing below "population."

A. *The Pithecanthropian leaf*

First, about 1890, we had the first *Pithecanthropus* (*P. erectus,*) an isolated enigma. Then, starting in 1930, the series of *Sinanthropus* in Northern China. Then, other remains of *P. erectus* in Java. And then, again in Java, the massive and brutish *P. robustus.* Then, still in Java, *Meganthropus*, with, in Southern China, another giant, *Gigantopithecus.* All these belong to the earlier Quaternary era. Meanwhile, moreover, we had (not properly understood at first, but later identified —as now seems unmistakably obvious—as a direct descendant of *Pithecanthropus*) *Homo soloensis*, of the later Quaternary in Java.

This is not the place to recount once again the detailed history and analysis of the successive finds that have suddenly, during the last twenty years, disclosed to us the number and variety of the types of fossil man that at one time were to be found along the Pacific edge of Asia. On the other hand, if I am to bring out what seems to me the true initial structure of the hominian group, I must emphasise the very remarkable (though too little appreciated) appearance of the evolutionary curve revealed in the distribution (at once geographical, and temporal and morphological) of these manifold witnesses to an extremely ancient human past.

We are always inclined to follow the easiest line and take too short and simplified a view of the developments of life. When it became established beyond doubt—particularly after the discoveries at Choukoutien—that *Pithecanthropus* was a true hominian, the first reaction of anthropologists was to imagine that in Trinil man and Peking man they had found, and could define in all his general characteristics, "Lower Quaternary man." It was the same illusion (though this already happened so long ago that we have forgotten it) that led so many excellent prehistorians, until about 1920, to think that all pre-glacial fossil men must be Neanderthal. The Sino-Malay evidence is now better known and better interpreted and can be studied as a whole, both on its own and in the light of recent African discoveries. We are, accordingly, beginning to think along quite different lines; that the fossil men of the Far East, so far from making us acquainted with an anatomical type that was "universal" for that period of time, represent in fact only a markedly differentiated (not to say almost detached) fragment of the true prehominians.

When we come to look into the question—and it becomes increasingly clear as we continue to do so—all the indications agree, in fact, in forcing us to accept this conclusion. There is the selective dissemination of *Pithecanthropus* along a well defined coastal strip that reaches out northwards (as far as Peking) from a clearly marked Malayan centre; there is the extreme variation we find of shape and size (the latter attaining the gigantic) within a highly determined osteological type (insignificant cranial convolution around the bi-auricular axis of the skull, powerful development of the occipital lobe); and then, too, there is their persistence in keeping to the same morphological line until the probable extinction of the group (*Homo soloensis*).

In fact, taking them all together, these various indications cannot but suggest to our minds what I shall call the notion of the "zoological scale"; by this I mean a natural unit, sub-phyletic in order, defined by the following characteristics: well marked individuality (both in habitat and shape), low miscibility with other elements in the phylum, considerable mutative power initially, the ability to prolong itself greatly in a residual form.

This idea that there are zoological "scales" and hence a laminated structure in every phylum (and the human phylum in particular) does more than clarify for us the physiognomy of the *Pithecanthropus* group; it has the further advantage of providing us with a general method of division that can serve to sort out in a truly natural and genetical order the still confused mass of fossil man. In a single segment of a fir-cone or a single leaf of a globe artichoke, we can read the structural law of the complete fruit. Similarly once we have identified the *Pithecanthropus* leaf *as such*—once that is we have recognised that taken all together the Java and Peking Men form a "scale"—we are encouraged to look in other places for traces of other similar units and also, so far as possible, to determine the numerical order of these various enclosed interlocking sheaths and their respective distances, in relation to a more or less ideal axis.

Let us then consider for a moment where, in the present state of our palaeontological knowledge, such a procedure leads us.

B. *The other leaves*

What makes the Pithecanthropian leaf stand out so distinctly for us, seems to be the two-fold fact that it developed marginally, at the extreme edge of Eurasia and that at the same time it represents a particularly precocious and therefore "outside" leaf of humanity: these two types of eccentricity, moreover (the geographic and the morphological) being closely inter-dependent. An ancient group is always a group that has been driven back: that rule has always held good ever since life started to spread out over the continents.

Farther west, nearer the heart, that is, of the Pliocene anthropoid patch, the phenomenon, as one might expect, becomes more confused.

At the southern extremity of Africa, it is true, we can see the emergence of the Australopithecine branch astonishingly similar to the Pithecanthropian shoot, and perhaps belonging to the same general biota, entering the long road to hominisation: this again is a marginal group, enclosed, in a state of

Cheese Barm

20   Lambart A Butler

active mutation, and, to complete the analogy, one that also includes giant forms. However, even though we should probably have to include this South African scale in the florescence of the human species (either as an abortive trial, or as a first tentative sketch), there can be no question, however typical it may be, of considering it as already forming part of what I called, earlier, the infancy of man.

Even if it should be proved that they were plantigrade, the *Australopithecines* are probably too ancient (Pontian) and their brain still too small for it to be possible to regard them as having crossed the threshold into reflection.

We have to admit, in fact, that in the whole mass of the ancient world we still do not know of any human scale that is clearly defined and for any considerable length of time. On the other hand, that such scales undoubtedly existed seems to be conclusively indicated by such traces as Neanderthal Man and Rhodesian Man: and they, to anyone who knows how to look at them, are the exact equivalents, the one in Europe and the other in Africa, of *Homo soloensis*. That such scales should largely have disappeared is satisfactorily explained by their presumed nearness to the main centre of hominisation. This zone of active expansion should most probably be located at the centre of gravity of the "anthropoid patch"—somewhere, that is, in the African continent—and it is not surprising that in that axial zone the rapidity of human pulsations should have prevented the mutations that succeeded one another (particularly the oldest and least adaptable) from becoming isolated, accentuated, and stabilised. In just the same way as we may anticipate, by a reverse process, that when we at last—if we ever do—discover their bone remains, the makers of biface implements in Kenya, the Cape, and the Narbada Valley, will seem anatomically much closer to ourselves than we now imagine: in them, we have the central forms of the human kernel; and in them, therefore, the true ancestors of *Homo sapiens*, himself the embryo of all modern mankind.

c. *The over-all plan*

In the diagram on the facing page I have tried to represent symbolically the general trend followed by the hominian group, expressed in the "scale system"; somewhat like a series of simple bodies arranged not in linear series but classified according to their period. This overlapping arrangement provides a ready explanation of the coexistence at different points on the globe of marginal and archaic, simultaneously with axial and progressive, types (or even, which is more baffling, of the pre-existence of the latter to the former, as in the case of Steinheim Man and Neanderthal Man); moreover, the explanation harmonises perfectly with the general drift of the whole towards states of progressively increasing cerebralisation.

There can be no doubt, then, but that it is in the direction and along the lines of "overlapping wholes" that human palaeontology must work in future, if it wishes, as in chemistry, to introduce a natural and fruitful order into its discoveries. And there is all the less doubt, I may add, in that the way of dealing with the human phylum that is so obtained corresponds exactly to that necessarily adopted in analysing the past, in every domain, whenever our analysis has the opportunity of studying sufficiently closely any centre whatsoever of organic expansion. As a general picture, the diagram in fig. 5 may be taken equally well as expressing both the rise of humanity at its birth, and the gradual establishment of civilisation (chapter IV). And, what touches more directly still upon our subject, it could equally serve to bring out, in its main lines, the structure of any or every other sufficiently fresh zoological group. On two occasions in particular, in the course of my scientific career—the first with the Oligocene cynodontids of Europe, and the second with the Pontian

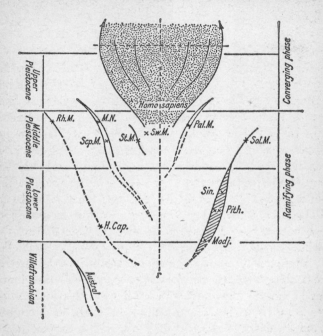

Fig. 5. The Hominian fascicle. Schematised structure on the "scale" hypothesis

Rh.M., Rhodesian Man; *M.N.*, Neanderthal Man; *St.M.*, Steinheim Man; *Sw.M.*, Swanscombe Man; *Pal.M.*, Palestine Man; *Scp.M.*, Saccopastore Man; *Sol.M.*, Solo Man; *Sin.*, Sinanthropus; *Pith.*, Pithecanthropus; *Modj.*, Modjokerto Man; *H.Cap.*, Homo capensis (Broom, 1949). *Austral.*, Australopithecines.

Note (1) The composition of the Pithecanthropian leaf, regarded here as supplying the structural key to the whole system; and (2) the folding back (or convolution) upon itself of the *sapiens* group under the influence of socialisation—a sort of "inflorescence".

mustelids of China—it has happened to me to meet with a fascicle of infant species. In each case, as any palaeontologist will readily appreciate, there was only one way of disentangling the complex I was studying, and that was to break it down into leaves, close-knit, rapidly mutating and with little to distinguish them from one another at the centre and the base—then, as they rose higher, spacing themselves out and fraying out into a small number of highly differentiated and stabilised types. We find exactly the same layout, whether we are dealing with men or carnivores, apart from a cardinal difference, as we shall shortly see, in the region of the kernel.

This leads us to the conclusion I wished to reach at the end of this first part: that the human "species," if observed as close as possible to its point of emergence, behaves essentially, in its beginnings, in exactly the same way as every other zoological phylum as it shoots into existence.

This does not mean—and I shall deal with this in my second part—that on closer inspection we do not find, even in the semi-embryonic stages of humanity, certain special properties of the highest importance which reveal the supra-specific, revolutionary character of the transition from instinctive life to reflection.

2. HOMINISATION, A MUTATION THAT, IN ITS DEVELOPMENT, DIFFERS FROM ALL THE OTHERS

Because we are men, and live among men, we end by being quite unable to see the phenomenon of man in its correct magnitude.

That remark will, indeed, be valid primarily for the two chapters that follow, when we shall deal with the "planetary" phases of hominisation. But it is already applicable at this point, in so far as, even though we have not yet directly encountered the great event of the socialisation of man, we are nevertheless already confronted by this surprising fact—that, from the end of the Tertiary period it is in man that the principal evolutionary effort of the earth is visibly concentrated.

The evidence is undeniable that, since the Pliocene, life seems to have concentrated in man (as a tree does in its leading shoot) all that was best in the sap it still held. In the course of the last two millions of years we can see that countless things disappeared, but not a single new thing, apart from the hominians, has appeared in nature. This significant fact should be enough in itself to demand our attention and awaken our suspicions. If, however, we now proceed to a more detailed analysis of the phenomenon, how are we to describe what we find? The vigour, the exuberance, the originality of this last-born of the children of earth! "A typical case of mutation": it was so that earlier we defined—labelled—the emergence of man at the heart of the "anthropoid spot" during the Pliocene. That, no doubt, is true enough, but only if we add "a mutation *unique of its sort,* inasmuch as almost from the very beginning there appear in the phylum to which it has given birth four properties, exceptional in their intensity and even quite unique in their novelty. These properties, which we must now examine in turn, are as follows:

An extraordinary power of expansion.

An extreme rapidity of differentiation.

A surprising persistence of germinative power.

And finally, a capacity, hitherto unknown in the history of life, for inter-connection between branches within a single fascicle.

A. *Extraordinary power of expansion*

Strictly speaking, it is only from proto-historic times (cf. chapter IV) that the astonishing power acquired by man of covering and possessing the earth becomes apparent and is given full rein. And yet, to an informed observer, the first indications of that power are already clearly discernible in prehistory. When we first meet man's tools and bone-remains at the beginning of the Quaternary, he is already occupying and even considerably overflowing (for example in Western Europe) the whole of the sub-tropical and tropical zone in which, from Africa to Malaysia, the evolution of the anthropoids had been carried through; and at the end of the period the great ethnico-cultural wave of the later Palaeolithic is spreading, with *Homo sapiens*, over the whole of the Old World, including the palaeoarctic zone. With this difference, that the connection between their branches was much looser, other phyla before man—elephants and horses, for example— showed themselves to be almost as irresistible invaders of the earth as man was; but none, among them all, seems to have launched its invasion along so wide and continuous a front, nor with anything like so vigorous a rhythm.

B. *Extreme rapidity of differentiation*

Here again—and this time not as regards his geographic extension, but if we look at his anatomical characteristics— man comes as a surprise to us when he emerges for the first time, already almost complete, into our field of vision. Whether we consider the dimensions of the brain, or the flattening of the face, or the specialisation of the lower limb, what a distance already separates the most primitive prehominians we know from, for example, the Australopithecines! Even making the most generous allowance for "mutation leap," such a gap can hardly be explained except by a particularly rapid evolution of the group in the first tens of thousands of years immediately subsequent to the first inception of hominisation. As the curve starts, we can only guess at the initial speed of transformation, but throughout the whole of the Quaternary period very distinct traces of it persist, in the human zoological group. No doubt (this is a point I touched on earlier, in chapter II, and shall have more than one occasion to return to), the fundamental difficulty we come up against in the study of evolution, once we reduce it (in the case of the "higher corpuscles" and most particularly of man) to a process of cephalisation, is that we have not yet succeeded in defining the essential factor and hence the *true*

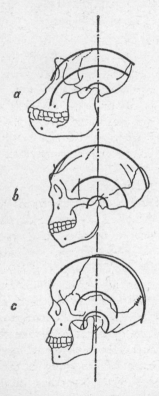

Fig 6. Cranial convolution in Man from the Anthropoids onwards (after Weidenreich). *a.* Gorilla; *b.* Sinanthropus: *c.* Modern Man

parameter of cerebralisation—quite apart from the fact that should we ever succeed in defining it scientifically, it will certainly prove to be a matter of neurones and not of osteology. This means that any attempt to measure on fossil skulls the progress of hominisation in terms of absolute value can at present be regarded as only roughly approximate. At the same time it is true that by judiciously using and combining certain external indications empirically associated with internal advances in nervous organisation (the increase in absolute volume and, still more, convolution of the cranium around its bi-auricular axis[1]—cf. fig. 6) we can follow in its main lines the development of the phenomenon. This is sufficient to justify the conclusion that between the moment when we see the hominians attaining the *Pithecanthropus* stage and when they *appear to us* to reach their zenith at the *sapiens* stage, they change, cerebrally, more rapidly and more profoundly than any other known living form during the same interval; even more, we may add, apparently, than the anthropoids themselves during the whole duration of the Miocene. So important a biological fact cannot, obviously, be overlooked.

c. *Persistence of phyletic germinative power*

By this I mean the remarkable capacity we find in the human type for an almost infinite production of new scales. In ordinary instances of zoological expansion the explosive phase of ramification that gives birth to a family of species is always short-lived. Thus, as I mentioned earlier (p. 45) (since it is impossible for us to note the very first phases of any "speciation") what we can apprehend in animal palaeon-

---

[1] This convolution results in an increase in the height and width of the brain case, disappearance of the occipital lobe and brow-ridge, and flattening of the facial area, this again producing in its turn the emergence of the chin, etc.

tology is never more than a fascicle of divergent trajectories radiating from and around an already "hollow" axial zone. In the case of man, however, it works differently. Let us look again at the diagram in fig. 5. Here we have a tentative grouping, according to their genetical and structural relationships, of the various human types so far identified by prehistorians. Had we been dealing with a rise of ruminants or carnivores, we should have had to expect, as I said, to see the centre of the sheaf weaken and empty as the Holocene approaches, so that at that level there would remain only a depleted corona of more or less isolated scales. What in fact we find, on the contrary, is that at just that very level—rising like a solid kernel in the heart of the axial region—the *Homo sapiens* fascicle makes its appearance, a witness to the vitality of a sap whose pressure seems to rise rather than drop with the passage of time. And I use the word *fascicle* advisedly, for the more closely one examines at this period the ultra-complex zoological system that extends to-day into modern man, the more one is convinced that it corresponds, anatomically, to an intense proliferation, a dense profusion, of scales (white, yellow, black and countless others, maybe): their lack of complete separation is evidence not, as an objector might maintain, of some inability to become fully individualised but (what is quite a different matter and inexhaustibly fruitful in its consequences) of the first expression of a quite remarkable force as yet unknown in Nature's history: the power of association and constructive agglutination among different leaves of the same zoological whole.

D. *Coalescence of branches*

Although the infra-human phyla are constrained to develop packed close together on the closed surface of the earth, they give evidence of no special aptitude for fusing into one another. Until man (and one could even say "until the prehominians," who also *seem externally* to obey the common law) animal evolution had functioned under the aegis of divergence. This accounts for the diffuse and overlapping structure that is so apparent in the tree of life (cf. figs. 2 and 5) from the largest boughs to the smallest twigs. Under the recognisable influence of the neo-centre of psychic attraction and inter-connection gradually created within the biosphere by the rise of reflection, it is the same system of dissociating differentiation that we see coming to a close at and beyond the level of *Homo sapiens*. *Homo sapiens* is an exasperating group for the classifier, because he no longer knows where, in this labyrinth of subtle, intricate anatomical characteristics, he should draw his lines of division; for the student of anthropogenesis, on the other hand, it is a group of compelling interest in as much as it is there that for the first time we can already clearly distinguish the functioning of a mechanism whose operation explains (as we shall have to show) the enormous lead over the rest of life built up by mankind in several hundreds of thousands of years. I am referring to the super-imposition, in biological evolution, of *convergence* on divergence, in such a way as to bring about a true organic synthesis of the potential species continually produced by phyletic ramification.

In that remarkable association constituted, about the middle of the Quaternary, by the *concrescence* of the most central or axial of the human "scales," in *Homo sapiens* that is, so far from meeting the last efforts of an exhausted evolutionary force, we hold the very germ from which sprang the definitive expression of the whole mass that lives and reflects. And, what is of even more importance, with *Homo sapiens* we leave the half-dark of mankind's infancy to attain a clear vision of the phenomenon of man, now at last recognised and defined as the establishment on our planet of a "noosphere."

CHAPTER IV

THE FORMATION OF THE NOOSPHERE

I. THE SOCIALISATION OF EXPANSION: CIVILISATION AND INDIVIDUATION

INTRODUCTION: PRELIMINARY REMARKS ON THE NOTION OF NOOSPHERE AND PLANETISATION

At the point we have now reached in our exposition, we may sum the situation up as it affects the world in process of corpuscular arrangement, as follows.

Thanks to the break-through of hominisation, the wave of complexity-consciousness on earth has penetrated, along the anthropoid phylum, into a domain or compartment that is completely new to the universe: the domain of reflection. And, once this barrier has been crossed, the wave (as in the past, whenever it has succeeded in breaking through one more "ceiling"), has again begun to split up into a complex fascicle of more or less divergent rays—the different radiating zoological lines of the human group. Since, however, as we saw at the end of the last chapter, these radiations are now propagated in a psychically convergent milieu, they soon showed a marked tendency to come together and fuse with one another. And thus was born, in an atmosphere of socialisation, if not as a result of it, the eminently progressive group of *Homo sapiens*.

Everything goes to show that socialisation (or the association in symbiosis, subject to psychic interconnections, of corpuscles that are histologically independent and strongly individualised), is an expression of a primary, universal, property of vitalised matter.[1] For a convincing proof of this all one needs to do is to observe how much each animal lineage, once it has attained its own specific maturity, demonstrates (in proportion to and according to the particular modalities of its own "type of instinct") the emergence of a tendency to group a smaller or greater number of its constituent elements into supra-individual complexes. At these pre-reflective levels, however (and particularly with insects) the ray of socialisation, however advanced that state may be, is still extremely weak, stopping short, for example, at the family group. It is true to say, then, that with man a new chapter opens for zoology, since for the first time in the history of life it is no longer a matter of a few isolated leaves: we now see a complete phylum—and, what matters even more, an ubiquitous phylum—suddenly and as one whole, giving evidence of becoming totalised: man, who appeared as no more than a species, but who, through the operation of ethnico-social unification, has gradually been raised to the position of constituting a specifically new envelope to the earth. He is more than a branch, more even than a kingdom; he is nothing less than a "sphere"—the noosphere (or thinking sphere) superimposed upon, and coextensive with (but in so many ways more close-knit and homogeneous) the biosphere.[2]

This and the chapter that follows will be devoted entirely to a study of the development of this new unity, planetary in dimensions, and of its properties: the proposition accepted

[1] As can already be recognised, at lower levels of autonomy in the constituent elements, in the formation of animal colonies (polyparies, etc.) or even in the metazoa (associated cells).

[2] To express the true position of man in the biosphere, we should need in fact a more "natural" classification than that worked out by present-day taxonomy. In the latter the human group appears logically only as a wretched marginal sub-division (family), whereas functionally it behaves as the unique, terminal, "inflorescence" on the tree of life.

initially and justified as we proceed being that if socialisation (as is proved by its "psychogenic" effects) is in every instance nothing more nor less than a higher effect of corpusculisation, then the noosphere, which is the final and supreme product in man of the forces of social ties, can take on full and final significance only if one condition be satisfied. That condition is that we look on the noosphere, taken in its global totality, as constituting one vast corpuscle in which, after more than six hundred million years, the biospheric effort towards cerebralisation attains its objective.

At the same time, I must hasten to add that the magnitude of this situation cannot be perceived all at once, nor was it so achieved. In its historical reality the planetary convolution of mankind upon itself proceeded only slowly: looking at it as a whole we may even say that it falls naturally into two major phases that it is important to distinguish with care. Supposing we imagine, inside a solid comparable to our terrestrial globe, a wave starting from the South Pole and rising up towards the North Pole. Taken over its *whole course* the wave in question advances in a curved and therefore "converging" medium: and yet at the same time, during the first half of its passage (as far, that is, as the Equator) it is spreading outwards; beyond that point, however, it begins to contract upon itself. Well, then: if we follow the historical development of the noosphere, we may truthfully say that it seems to conform to an exactly similar rhythm. From its origin until our own time, mankind, while gathering itself together and already in the first stages of organisation centred upon itself[1] certainly went through a period of geographical expan-

---

[1] This, I must emphasise, is something that none of the phyla that had appeared earlier in the biosphere, however ubiquitous they may have been, had yet succeeded in doing: this was in spite of their being compressed on the closed surface of the earth, and is explained by their lack of an appropriate psychism.

sion, during which its first concern was to multiply and inhabit the earth. And it is only quite recently that, "once the frontier was crossed," the first symptoms appeared in the world of a definitive, global, folding back upon itself of the thinking mass within a higher hemisphere: and once that has been entered, it can, under the influence of time, advance only by contracting and concentrating itself.

Thus we find a reversal of the socialisation of expansion, to culminate in socialisation of compression.

In this chapter, let us confine ourselves to a study of the first, only, of these two phases, reducing its vicissitudes or characteristics to the three following heads: populating, civilisation, and individuation.

I. THE POPULATION OF THE WORLD

In the human zoological group, the remarkable power of expansion that characterises it (see above, Ch. III) is obviously linked with advances in socialisation. It is because it became capable, through its attainment of reflection, of assembling and buttressing together indefinitely its constituent elements, that mankind, the last-born of evolution, was able so rapidly to establish itself throughout, and ultimately above, all the rest of the biosphere. In such circumstances it is not surprising that when we now look back on the populating of the world it seems to us to have been brought about in a succession of ever widening pulsations: each new pulsation corresponding to a new and better social arrangement of the hominised mass.

In the axial (Mediterraneo-African) zone of hominisation— where, that is, the successive human waves are superimposed in too close succession and over too long a period of time to be readily distinguished—in that zone the rhythm and different phases of this stop-and-go invasion are still obscure. On the other hand, when we look at a vast marginal area, such as Eastern Asia, where each new wave could, as it started, find enough room comfortably to overrun its predecessors, three major overrunnings at least (very broadly speaking) are now seen to stand out. The first two (noted here simply as a reminder) belong to prehistoric times, but the third definitely initiates the modern historic regime of the expansion of man.

First pulsation: the pre-hominian wave, running from south to north along the Pacific coast. We can say practically nothing about the cultural level of this primitive humanity: except that at Choukoutien (at the extreme limit of the wave, that is)[1] *Sinanthropus,* who used fire and worked stones, gives the impression of having belonged to a group that was already appreciably socialised: and that, no doubt, is just what explains the remarkable power of expansion and ethnic penetration that was able to carry him from the subtropical zones of Asia as far as the first escarpment of the Mongolian plateau.

Second pulsation: the Aurignacian wave of the later Palaeolithic: advancing from west to east, and particularly well marked in the loess areas of the Yellow River. I referred earlier (chapter III) to this exceptionally powerful wave, thrown up by the coalescence and emergence of the *sapiens* group: it brought with it not only fire but art, and its deposits (immediately recognisable by their elaborate bone and stone industry) extend over practically the whole of the old world. In the axial or southern regions of the globe, on the one hand, they cover, in sharp contrast, the old palaeolithic levels; on the other hand, in what had until then remained a palaeoarctic no-man's-land, they are spread out from west to east over a virgin soil, from the northern Alps to the Pacific.

[1] And on the hypothesis (which is by far the most probable) that Peking Man is indeed the originator of the industry found in association with the bone remains in the archaeological deposits.

Third pulsation: *the neolithic agricultural wave.* Towards the end of the Pleistocene, the slow cumulative action of closer ethnic ties and cultural exchanges brought about a decisive change within the *sapiens* fascicle, which was now (as a result of the gradual disappearance around it of all the other pre-hominian scales) the only one in which the future of hominisation on earth was to be achieved. Practically everywhere in the area that had been populated in earlier periods, but particularly along two wide strips—one North African or Mediterranean, the other North European and Siberian—there are numerous indications, about this time, of a more sedentary and more fully grouped way of life. These signs herald the great neolithic metamorphosis in which (simultaneously, it appears, over wide areas) mankind passed (under the influence of some sort of generalised maturation) from a diffuse to an organised society. This was principally due to the discovery of agriculture and stock-raising; for these are forms of industry that not only allow but demand a rapidly increasing demographic density and internal organisation among the populations engaged in them.

This transformation was already well marked in the period known as "mesolithic" (about ten or fifteen thousand years before the Christian era) and its effect was to cause a sharp rise of human pressure in the areas affected by it: under its influence a new ethnic surge, more powerful than any of its predecessors, made itself universally felt, being most particularly clearly marked in the Siberian strip. Here a migratory mass was built up, that was able not only to overflow, south of the Altai, as far as the Yellow River country ("Mongolian" neolithic),[1] but even to reach Alaska (just at that time ice-free) and, once it had established that bridge-

[1] Cf. P. Teilhard de Chardin and W. C. Pei, *Le Néolithique de la Chine* (Publications of the Peking Geobiological Institute, no. 10, 1944).

head, to move on and occupy the two Americas from end to end.[1]

It was at that moment that we might say the first features of the Noosphere were definitively drawn: but that was still only in an embryonic and tentative manner. On the other hand, when mankind reached the extremities of the New World it was certainly quite unconscious of having completed its own circle. The network, again, that was woven in the course of this supreme advance was still so slack in its "fabric," and so heterogeneous in its threads that no influence could, obviously, still be transmitted through it, except with extreme slowness, dispersion and wastage.

This fragile membrane had to be consolidated and built up into a solid structure, either by organising on the spot groups that were already installed, or by the periodic influx of new elements. This now emerges as the great task of civilisation.

2. CIVILISATION

A *The biological nature of the phenomenon*

History is at length leaving behind a long descriptive phase during which its chief concern was an exact and colourful reconstruction of the past; it is now tending more and more to offer itself as a science of the *laws* that underlie the apparent capriciousness of human vicissitudes. To appreciate the true character of this new organically inspired outlook, it is sufficient to turn to Arnold Toynbee's monumental work: in this he first lists twenty-one distinct civilisations from Sumerian and Minoan times to our own day, and then concentrates on dis-

---

[1] This operation must have taken thousands of years, since if the migratory peoples were to advance they had to create a new type of agriculture at each new latitude; at the same time we must presume that it was completed early enough for the domestication of plants to have been completed even in South America (manioc) well before the arrivals from Europe.

tinguishing in them the conditions of their birth in different geographical surroundings,[1] the mechanism of their growth,[2] their reactions upon one another and their decline, the rhythm in which they succeeded one another, etc.[3]

An attempt of this nature, and one so massive, brings out very clearly the irresistible tide that for the last hundred years has been bringing natural history and human history closer together. Even so, the basic concurrence of the two disciplines is still far from appearing complete—nor, indeed, is it even clearly envisaged. In both Toynbee and Spengler the social evolution of man is, no doubt, treated *in a biological way*—but it does not thereby cease to remain outside and separate from biology. The domain of zoology and the domain of culture: they are still two compartments, mysteriously alike, maybe, in their laws and arrangement, but nevertheless two different worlds. The most organically aware of historians seem definitely to have halted at that dualism—without, moreover, any apparent surprise or uneasiness.

Now: it is at this point and this particular juncture that the view adopted here of a universe in process of general involution upon itself comes in as an extremely simple way of getting past the dead end at which history is still held up, and of pushing further towards a more homogeneous and coherent view of the past. There is no difficulty in this once we see what civilisation, expressed in terms of its biological mechanism,

[1] The fluvial type (Egypt, Sumeria, the Indus . . .); the plateau type (Andean, Hittite, Mexican civilisations . . .); the archipelagic type (Minoan, Hellenic, Japanese civilisations . . .).

[2] A growth that operates principally under the stimulus of having to meet problems of survival presented by environment (the "Challenge and Response" theory).

[3] A rhythm punctuated by the periodic formation of "universal empires," the fall of each stimulating the launching of a new ethnic wave and the appearance of some "universal religion."

really amounts to. By civilisation I mean not a fully realised state of social organisation but the actual process that generates the organisation, and in that sense civilisation is, ultimately, simply zoological "specialisation" extended to an animal group (man) in which one particular influence (the psychic) that had hitherto been negligible from the point of view of taxonomy suddenly begins to assume a predominant part in the ramification of the phylum. It is the same thing, but on a new plane. Indeed, we are, and have long been, perfectly familiar with any number of animals (among insects, for example, birds and rodents) whose instinctive behaviour provides the classifier with differentiating characteristics at least as well marked as colour, size and shape. It seems only reasonable, then, to generalise and push to its limit this notion of "psychological species", and so recognise and admit that the multiple and multiform human "collective units" produced in the course of history as a combined effect of culture and race are, in the domain of the reflective and free, groups just as *natural* as any variety you please of ruminant or carnivore. There is only this difference to be allowed for, that in this case the psychic plays a more important part than the physiological and morphological, and so certain properties or "liberties," of a type hitherto exceptional or even unknown, appear in the operation of vital forces. First among these is that, since the older chromosomic heredity is now partnered by an "educational," extra-individual, heredity, the preservation and accumulation of the *acquired* suddenly assumes an importance in biogenesis of the first order.

From this point of view the formation of tribes, nations, empires, and finally of the modern state, is simply a prolongation (with the assistance of a number of supplementary factors) of the mechanism which produced animal species; and thus the history of man for three reasons among others, is seen to be a specially favourable field for the study of the laws of phylogenesis. These reasons derive in the first place from proximity of association—we might even say "interiority"—since the evolutionary phenomena that make up that history are not only all compressed within the span of the last thousands of years but are still continuing in, and are central to, what we are experiencing at this very moment. Another reason is their sharply defined character, inasmuch as the various strands that appeared in succession as the noosphere extended itself (coloured as each of them is in the bold tints characteristic of one particular cultural complex) are easier to follow and distinguish on the whole than the purely anatomical elements in any one zoological group. This is so much the case that in the last resort we shall do best to rely primarily on the biology of civilisations if we wish to check, determine more accurately and confirm in detail (as in a well laid out specimen) the rough general picture that palaeontology has already given us of the great evolutionary laws of orthogenesis and differentiation.

B. *Effects of differentiation*

Once we have raised the completely artificial barrier that still (by habit or convention) separates the two processes of socialisation and vitalisation, we immediately find that a basic simplicity (the same as that we have already met in the pre-reflective zones of the biosphere) can be distinguished beneath the apparent irregularities and disorder of the human adventure. The birth, the migrations, the conflicts, the substitution one by another of a hundred different peoples—what, when you finally analyse it, is all this polymorphous, motley effervescence, what is it fundamentally, if not the operation, endless and never changing, of the ramification of living forms continuing to function in a civilised context?

Initially, we have the "basic" skein of the great races (white, black, Mongoloid) that emerged from the Pleistocene. And then, starting from this primordial ethnico-cultural fascicle we find again, periodically, "in pulsations," the formation of new scales, new rays that branch off, exactly similar in their behaviour to any other zoological scales or rays: with exactly the same way (and for the same reasons) of emerging suddenly, already practically ready-made, on the horizon of history[1]; with the same way of becoming fixed and set, more or less rapidly, in a secondary state of immobility; with the same tendency to fade away as they are relieved by some neighbouring ray, itself in turn born, we hardly know where, of some embryogenesis we cannot determine.

All this, I must emphasise, is an admirable verification and confirmation (within a system—the human social group—whose perfect monophyletism, in spite of any gaps, no one can deny) of the general laws of animal phylogenesis. At the same time, however, all this develops within an enriched and rejuvenated biological atmosphere in which, as a result of the intensification of the psychic milieu, a phenomenon hitherto unknown in nature has now become possible—the *confluence* of branches. Within the pre-human biosphere the distribution of living forms could be followed and explained in terms of appearances and disappearances, that is, simply by the operation of external forces and resistances among the living groups in question. In the case of human aggregates, on the contrary, whose interaction on one another has come *from within*, a new regime is inaugurated: in this, besides the elementary operations of penetration, elimination and substitution, allowance has to be made for the much more complex phenomena of interphyletic combinations. And this has, among other consequences, the two following.

The first is that we now have to take into account a hitherto

[1] We know as little about the origins of the Greeks or the Chinese as we do about that of the mammals or amphibians.

unknown and peculiarly revolutionary type of mutation: this is no longer the result of a re-arrangement of the germinal particles within certain individuals, but of the massive cross-fertilisation of large ethnic groups suddenly brought together as chance governs their migrations or expansion. It was thus, without doubt, that the first kernel of the Mediterranean civilisations was formed; and thus, too, that in Alexander's time the world began to attain a real awareness of its unity when, as Grousset puts it,[1] the three civilised mankinds of that time (Greece, India and China) suddenly realised that they inhabited the same planet; and it was thus, finally, that by the successive discoveries of America and Oceania the West took over (and for a long time still, it seems) the direction of human destinies.

The second consequence is that our attention is drawn, or rather forced, once again to the orientated, "orthogenetic" nature of an evolution whose controlled character—which, in the field of pure morphology, is just possibly contestable—is undeniably evident in the field of science—if only when we note how the mosaic of neolithic tribes was able, by the conquest, fusion and progressive articulation of its elements, to produce the map of modern nations or states as we find it in our atlases.

c. *Effects of orthogenesis*

By orthogenesis (in the widest and most strictly etymological sense of the word) we should in this context, I repeat, understand the fundamental drift as a result of which the stuff of the universe is seen to behave as though moving towards corpuscular states continually more complex in their material arrangement and, psychologically, continually more interiorised: this drift, we should add, being, in the case of

[1] Cf. R. Grousset, *De la Grèce à la Chine* (Monaco, "*Les Document d'Art,*" 1948), p. xi.

higher living beings, directly involved in an increasing concentration of the nervous system.

In fact, throughout the duration of the historical periods occupied by what I called earlier the expansion phase of socialisation, it does not appear possible, at least for the moment, anatomically speaking, to record any particularly marked advance in the structure of the human brain. Whereas, during the Quaternary, a quite appreciable progress in the convolution and convexity of the brain-case can be noted from the prehominians to *Homo sapiens*, we find nothing (except perhaps, if we are to believe Weidenreich, a certain general tendency to brachycephaly) from the end of the Palaeolithic, throughout the last twenty thousand years, that noticeably indicates any new step forward in cephalisation. So much so, indeed, that students have often been inclined to conclude that, in man, cerebralisation is reaching its zenith in this quasi-stationary state[1]—if, indeed, it has not been halted completely.

Now, this is to overlook the appearance in nature, with man—precisely through this wonderful device of socialisation in a reflective milieu—of a new type of "psychogenic"[2] arrangement (educational and collective in nature—cf. p. 87 above) that came in at just the right moment to reinforce, or relieve[3] the older and perhaps exhausted forms of cerebralisation.

Supposing we admit provisionally, and subject to any necessary reservation that, in its histological arrangement, the

[1] This may simply appear to be so, owing either to the shortness of the period in question (twenty thousand years is nothing for even an accelerated biological evolution) or to its being impossible for us (as pointed out in chapter IV) to distinguish, behind a number of crude osteological details, the subtle and as yet ill-understood influence of the organisation and arrangement of the neurones.

[2] "Psychogenic" in the active sense of "generating consciousness."

[3] Or even to give a new upward impetus to (cf. chapter V).

individual human brain has, since the end of the Quaternary, really arrived at the limit set by physics and chemistry to its progress in complexity: even then it would still remain true that since that time, as a result of the combined, selective and cumulative operation of their numerical magnitude, the human centres have never ceased to weave in and around themselves a continually more complex and closer-knit web of mental interrelations, orientations and habits just as tenacious and indestructible as our hereditary flesh and bone conformation. Under the influence of countless accumulated and compared experiences, an acquired human psychism is continually being built up, and within this we are born, we live and we grow—generally without even suspecting how much this common way of feeling and seeing is nothing but a vast, collective past, collectively organised.

To anyone whose mind is sufficiently sensitive to the appreciation of such biological realities of a higher order, nothing can be more obvious than that in the two-fold phenomenon of man's conquest and organisation of the earth we see a direct extension of cosmic convolution. Indeed, the really important point to be decided is already no longer whether the current of hominisation around us may not perhaps be slowing down: this is because with, and following upon, the coming into play of the effects of civilisation, anthropogenesis has now really got into its stride. No, all we have to settle now is towards what sort of biological fulfilment the irresistible forces of orthogenesis, in their rejuvenated form, are leading us.

And this brings us to the consideration—though it is still open to us to reject and go beyond it—of the solution, already so popular in spite of its shortcomings and harmfulness, of *individuation*.

3. INDIVIDUATION

Precisely because of its essential mechanism (which consists in a "chain corpusculisation"—cf. chapter 1), the phylogenesis of living forms can be carried on only at the cost of a permanent and continually fiercer conflict between lineage and individual, between present and future. So long, throughout an animal series, as the independence of the successive somata is still so limited that the latter remain true, by and large, to their role as links, the phylum develops normally, protected and consolidated internally by a vigorous "sense of the species." But as the elements of the phyletic chain, as a direct consequence of the advances in corpusculisation, increase in interiority and liberty, so the "temptation" inevitably grows stronger for them each to set itself up as the term or head of a species and to "decide" that the time has come when each must now live for itself.

A jet of water that breaks up into droplets as it approaches the top of its flight—it is in just that way that the phenomenon of the "granulation of phyla" presents itself to our experience. It is a phenomenon that can hardly be distinguished in the domain of pre-reflective life, but one destined to take on a rapidly increasing importance in the case of man, and above all of socialised man. According to the most skilled observers,[1] a sort of collective co-consciousness can still be found among tribes classified by ethnologists as "primitive," which in a quite natural way facilitates the cohesion and proper functioning of the group. It must have been so practically everywhere

[1] Cf., for example, B. Malinowski, *Argonauts of the West Pacific*, which describes the *Kula*, an intricate and complex magico-commercial organisation, that functions annually without any of the participants appearing to have a clear appreciation of the process as a whole. See also Gerald Heard, *The Ascent of Humanity* ("from group-consciousness, through individuality, to super-consciousness").

on the earth in pre-neolithic times. On the other hand, as civilisation began to rise, so an increasing agitation became continually apparent in populations each consistent element of which felt itself under pressure from a more lively capacity for, and hence need of, autonomous activity and enjoyment. This was so marked that towards the end of the nineteenth century it could seriously be asked whether hominisation was not approaching, through pulverisation and fragmentation, its final phase.

At that time, in fact—corresponding historically to the full "expansional" deployment of the noosphere—the isolation one from another of human particles, their self-centred tendencies now heightened by the first establishment of a practically universal culture, reached, as one might expect, its maximum. At the same time the "sense of the species" was automatically (as a result of an internal slackening) dropping to its minimum, within a phylum whose layers were spreading out so uncontrollably as to cover the whole earth. This was the age of the rights of man (i.e. of the "citizen") against the community: the age of democracy, naively conceived as a system in which everything is for the individual and the individual is everything: the age of the superman, envisaged and awaited as standing out in isolation above the common herd.

All these various converging indications prompted the belief (still held, it seems, by many?) that mankind, like a liquid that has started to boil, has arrived at some limiting and critical state of organisation, and that there now lies ahead of it no biological possibility nor destiny other than to generate (and release in a state of isolation) particles that are increasingly more self-sufficient and self-centred.

It is only fifty years ago that civilisation reached a sort of paroxysm in the West and gave every indication of being about to culminate in *separate* persons, that is in individuation.

94

It was, however, at that very moment that there began to appear over the horizon, like clouds charged at once with storms and the promise of good things to come, the massive and as yet undreamt of forces of totalisation.

CHAPTER V

THE FORMATION OF THE NOOSPHERE

II. THE SOCIALISATION OF COMPRESSION: TOTALISATION AND PERSONALISATION: FUTURE TENDENCIES

I. AN ACCOMPLISHED FACT: THE INCOERCIBLE TOTALISATION OF MAN AND ITS MECHANISM

With our eyes still dazzled by the prospects (or rather, for reasons we shall soon appreciate, the *mirage*) briefly disclosed to us by modern doctrines of individuation, we continue more often than not, in the middle of this twentieth century, to dream of a world in which every man would find in the progress of his social environment simply a continually more effective jumping-off ground from which to reach a way out in a completely independent and "individualistic" solution of the problem of life. It is a prospect as pluralist as a shower of sparks, in which, in each individual case, the limit of the world is identified with the end of each reflective element: the element being considered on its own, in the incommunicable solitude of what separates it from all the others. And because our eyes are enslaved by a sort of firework display that gives us the illusion that plenitude awaits us, our attention turns away with indifference or irritation from another and quite different possibility: and yet in every field, economic, political, and philosophical, the signs that herald this latter are manifold, warning us that socialisation, far from becoming comfortably domesticated (as we flattered ourselves) for our own private use, is in fact pushing forward even more vigorously, following an irrepressible process of unification whose mechanism, operating plainly for all to see, is governed by three well marked periods, as follows:

A. *First period: ethnic compression*

Here we meet (experientially speaking) the mainspring or initial motive force of the whole phenomenon. As we all know from our own experience, the human population is coming close to saturation point on the closed surface of our planet; it is thus, through the internal forces of reproduction and multiplication, becoming continually more compressed, and the effect of this compression is to create, at the heart of the noosphere, a constant or even increasing source of available energy. If it were simply some gaseous mass that was involved in a process of this nature, such a proliferation of particles would make itself felt in some mechanical or thermal effect: there would be an increase of heat or pressure. In the case of human (or, more widely, living) corpuscles, there is a more subtle transformation of energy. It is expressed ultimately no longer in a simple numerical equivalence but in an effect that takes the form of arrangement. Hence:

B. *The second period: economico-technical organisation*

Compress some inanimate matter, and you will see it react, in order to avoid or respond to your action, by a change of structure or state. But compress some vitalised matter (subject, of course, to determined precautions and within determined limits), and you will see it organise itself. There is, perhaps, no more universal law than this, to explain the genesis of the biosphere and still more of the noosphere. Without the reciprocal pressure of corpuscles (that is, supposing they existed in a space that was completely elastic or completely without tension) life would probably never have appeared in the world: still less would reflection—nor, *a fortiori*, human society. And, conversely, civilisation could have reached the pitch and level we see to-day—and which makes us realise the mysterious relation between hominisation, the force of gravity, the surface area of the continents, and the radius of the earth—only because of a certain optimum ratio between the dimensions of our being and the curvature of the heavenly body that bears us. To see how true this is, one has only to look at the two comparative curves of culture and demography. Particularly since the Neolithic, the more mankind is compressed upon itself by the effect of growth, the more, if it is to find room for itself, is it vitally forced to find continually new ways of arranging its elements in the way that is most economical of energy and space. This has the most remarkable result (though a biologist might well anticipate it) that, under the stimulus of this need and inspired by this search—as a result, too, of the new devices that are contrived —what appeared at first no more than a mechanical tension and a quasi-geometric re-arrangement imposed on the human mass, now takes the form of a rise in interiority and liberty within a whole made up of reflective particles that are now more harmoniously interrelated. And this brings us to the third period in the operation.

c. *The third period: simultaneous increase of consciousness, science and radius of activity*

It is not, in itself, surprising that a rise in "psychic temperature" should automatically accompany a better social arrangement. It is simply another instance of the fundamental law of complexity-consciousness that runs through the whole of this book and guides our investigation. On the other hand, we come to the really interesting point when we realise that this increase in mental interiority and hence of inventive power (in which man's compression upon our planet is ultimately expressed) simultaneously and inevitably increases each human element's radius of action and power of penetration in relation to all the others[1]; and, in proportion as it does so, it has as its direct effect a super-compression upon itself of the noosphere. This super-compression, in turn, automatically produces a super-organisation, and that again a super-"consciousisation": that, in turn, is followed by super-super-compression, and so the process continues. The cycle follows an organically welded chain that completes the circle: but, what is more, it continues indefinitely to build up its own intensity, as happens with a properly tuned amplifying system. The process is so marked that anyone who takes the trouble to analyse, as we have just done, the mechanism of the economico-technico-social forces whose network has now been spreading insidiously over the world for a century, will find it unmistakably evident that it is absolutely impossible for us to escape the forces that draw us together: in the preindustrial periods of history their uncontrollable pressure increased almost unnoticed; to-day we see it brought suddenly into the open in all its strength.

Quite apart from any scientific or philosophical presumption, and without anticipating any judgment of value, we are now obliged to face a situation—or rather a generally experienced condition—that in fact demands recognition as objectively and implacably as the movement of the heavenly bodies or the decomposition of radio-active substances; and it would be completely useless, in any domain, to try to build up anything that could stand out against that condition.

"Thus through the combined influence of two curves, both cosmic in nature—one physical (the roundness of the earth)

[1] To-day, thanks to the single discovery of electro-magnetic waves, any man can immediately and simultaneously make contact, through what is most human in himself, with all men on earth.

and one psychic (the reflective's self-attraction),[1] Mankind is now caught up, as though in a train of gears, at the heart of a continually accelerating vortex of self-totalisation."

There we have the brutal fact that we must now try to understand.

2. THE ONLY COHERENT EXPLANATION OF THE PHENOMENON: A CONVERGENT WORLD

When we see that our attempts to break through the circle that binds us together are constantly checked, and it finally becomes clear that the forces of compression that invest us may well be no temporary accident but the symptom and first indication of a permanent regime now being stabilised in the world we live in, and that it has come to stay—then, a truly "mortal" fear tends to possess us: the fear that, in the course of the transformation we see heralded, we may lose the precious spark of thought, so painfully lit after millions of years of effort—our own little "ego": the essential fear of the reflective element when it faces an apparently blind whole, whose vast layers enfold it as though to re-absorb it while still in the fullness of life. . . . Have we, we ask, emerged into consciousness, and not only into consciousness but (as Lachelier says) into consciousness of consciousness only to sink back immediately into an even blacker unconsciousness?—as though life, having carried us at arm's length into the light, then fell back exhausted?

On first consideration, this idea, depressingly pessimistic though it be, of a decline or ageing of the spirit through a general anchylosis of the human mass, has some appearance of truth. The first effects, entailing unmistakable slavery, of

[1] Only the first of these two curves has any appreciable action on pre-human life: hence the impossibility for the biosphere (as opposed to the noosphere) of centring upon itself.

factory work; the first forms, brutally herding men together, assumed by political state control; the terrifying example (all the more terrifying because ill-understood)[1] of ants and termites—all these impressive symptoms justify, up to a point, the instinctive reaction of apprehension and recoil that, as we can see for ourselves, forces so many human beings in desperation, when faced by the inexorably rising pressure of the noosphere, to take refuge in what are now obsolete forms of individualism and nationalism.

It is here that it becomes essential, if we are correctly to understand what is going on, to proceed scientifically. By this I mean that at this particular juncture we must re-plot on as wide a trajectory as possible, the highly critical section of the curve in which we are at present living. We must, therefore, look at it from further back and higher up; and to do this we must return to the point of view of a universe in process of involution. If we do so (and such a view-point has so far been a sure guide to us throughout this inquiry) we shall most certainly find that our fears of "dehumanisation by planetisation" are exaggerated; for the planetisation we so dread is simply, to judge from its effects, the authentic, direct, continuation of the evolutionary process from which, historically, the human zoological group emerged. We were noting only a moment ago that the final result of the physico-social compression to which we are now being subjected is a rise in the psychic temperature of the human mass. Anyone, then, who has followed our earlier argument, will need no further proof of the true nature of the type of super-grouping towards which we are being driven by the continual development of civilisation: it is not just another of those material aggregations ("pseudo-complexes") in which the various

[1] By which I mean that allowance is not made for the radical difference between the "mechanisable" psychisms of insects and the "unanimisable" psychism of man.

elements' liberties are either cancelled out by the statistical effect of large numbers or, by geometric repetition, become simply mechanical. On the contrary, this super-grouping is one of the type of "eu-complexes" (see chapter I) in which the arrangement, because it is, and in as much as it is, productive of consciousness, is *ipso facto* to be classed as biological in nature and value.

In fact, if we watch what we are doing, we shall find in the current of totalisation that at present seems to be trying to snatch us away from ourselves and de-centre us, simply a fresh beginning—still the same, but on a higher plane—of the process of corpusculisation that generates life. After appearing to have reached its zenith in producing the seed of reflective consciousness, this same process is now setting about grouping together and synthesising these seeds of thought. After man, we get mankind . . . a movement, as we know, adumbrated ever since the pre-hominids, pursued in a subtly and secretly pervasive form throughout the growth of *Homo sapiens*, but only now, and that for a very precise reason, entering its critical phase of encirclement.

Here we may well return to the comparison (cf. chapter IV, p. 81) with which we began our study of the noosphere: the wave of hominisation advancing from the North to the South Pole within an imaginary globe. In this picture the modern crisis of individuation corresponds to the arrival of the wave at the equator. Here we have the extreme point of separation, which means of independence, between the highly differentiated elements, to be found during the expansion of civilisation. At the same time, however, the position is one of unstable equilibrium; with the earth demographically saturated, the least increase of mutual pressure among such highly charged human molecules was bound to bring about the reversal of which we are at once the agents, the objects, and the witnesses; the change from one hemisphere to another—the universe suddenly closing over our heads like a dome—the transition from expansion to compression.

In earlier days, man's consciousness could be revolutionised simply by the discovery of a new continent: if that was so, what, indeed, are we to say of the revolution that is now going on in our minds as a result of the appearance—fortunately it has been gradual, and gently broken to us—of the extraordinary realm into which the irresistible action of a contracting world is forcing us to enter and pursue our advance? Like a doctor at a sick-bed, we often wonder why all around us we should see this hitherto unknown amalgam of anxieties and hopes bringing restless uneasiness to individuals and to nations. Surely the basic cause of our distress must be sought precisely in the change of curve which is suddenly obliging us to move from a universe in which the divergence, and hence the spacing out, of the containing lines still seemed the most important feature, into another type of universe which, in pace with time, is rapidly folding-in upon itself.[1] This brings with it a radical structural and climatic change that at one blow deeply influences and re-shapes our whole outlook and activity. Since the sixteenth century man had learnt, in turn, that the cosmos in which he lived was in movement—and that the movement consisted primarily in an arrangement directed towards super-life. It is only now that he is taking the third, and most perilous, step and beginning to see that cosmogenesis, so defined, is not only going on over his head, but is tending to complete its circular motion much more rapidly than had been thought.

At this decisive moment when for the first time he (man, that is, man as such) is becoming scientifically aware of the general pattern of his future on earth, what he needs before anything else, perhaps, is to be quite certain, on cogent

[1] This "crossing of the equator" may perhaps explain the terrible political and social storms we are now living through.

experimental grounds that the sort of temporo-spatial dome (or cone) into which his destiny is leading him is not a blind alley where the earth's life-flow will shatter and stifle itself. Man is now realising that this cosmic spindle corresponds, on the contrary,[1] to the concentration upon itself of a force that is destined to find in the very heat released by its convergence sufficient strength to burst through all the barriers that lie ahead of it—whatever they may be.

3. EFFECTS OF, AND FORMS ASSUMED BY, CONVERGENCE

A. *Increase of free energy, and intensification of research*

When, a few pages earlier (pp. 96-100) we were analysing the chain-structure of the "economico-technico-scientifico-social" complex whose appearance is characteristic of a socialisation that has attained its "equatorial" point of reversal and compression, we pointed out that, by the very nature of its functioning, the system attracted our "liberties" towards progressively higher organico-psychic states. Considered under this aspect, the Noosphere, when in process of concentration towards the pole, behaves like a body that gives off radiation— the radiation being produced by a free energy, the nature and changing forms of which we must now briefly examine.

Initially, the free energy in question is nothing more nor less than the quantity of human activity (at once physical and psychic) made available by the two allied advances of social co-operation and *mechanical skills*. As I have had countless occasions to say again and again, nothing is more unfair or a greater waste of time than to protest and fight against the increasing leisure towards which the machine is inexorably leading us. Without the very many automatic processes whose business it is to make our various bodily organs work

---

[1] Provided, of course (cf. pp. 118-21), that the play of our liberties lends itself to it.

"on their own," none of us, it is obvious, would have any "leisure" to create, to love, or to think: the necessity to look after our "metabolism" would occupy us entirely. Similarly (and allowing, of course, for all the difficulties associated with the absorption of a too sudden release of man-power) we must realise that the continually more complete industrialisation of the earth is simply the humano-collective form of a universal process of vitalisation which, in this as in all the other cases, can only lead, if we know the right way in which to approach it, to interiorisation and freedom.

We are now faced by the torrents of unapplied power already released by the convergence (little advanced though the process be as yet) of the human mass. A too common reaction—and an absurd and unnatural way to behave it is—is to try to force back this disconcerting outburst: the right action to take is surely to direct the flood along the slope to which its natural inclination is clearly leading it—and by that I mean in the direction of research.

If we define research as an effort to feel our way towards the continual discovery of better biological arrangements, we may (and even must) agree that very generally speaking it represents one of the fundamental properties of living matter. Taking it now in a stricter sense, with its usual meaning of *reflective* groping, research, again, is necessarily as old as the awakening of thought on the earth. And yet, looked at in the generalised and conscious fullness of its operations, research (it is essential for us to realise) corresponds to an entirely recent and extremely significant development of hominisation.

In this case, as in so many others, I appreciate that the slowness of life's movements may well deceive us and blunt our perceptions. All we have to do, however, is to try to get a picture of mankind at two points sufficiently far apart in duration for the general drift of the system to become apparent. Or, better still, let us place ourselves successively at two points lying on either side of a particular phase in which there is a rapid change of direction: let us, that is to say, compare from the point of view that concerns us, the state of the world as it is at the present moment with that in which it was still to be found, for example, between the Renaissance and the French Revolution. From such a comparison two most illuminating pieces of evidence emerge.

The first is the sudden and enormous importance (both qualitative and quantitative) acquired, in less than two hundred years, by science and technology in the field of human activities. Almost until the beginning of the nineteenth century, as is common knowledge, the scientist was still, on the whole, the exceptional being, the "oddity," isolated from his fellows by his "hobby"[1] or dream: a type sporadically distributed throughout the human mass and but lightly connected with it. To-day we find the reverse: research students are numbered in hundreds of thousands—soon to be millions —and they are no longer distributed superficially and at random over the surface of the globe, but are functionally linked together in a vast organic system that will remain in future indispensable to the life of the community.

The second piece of evidence is the impressive coincidence of the decisive establishment on earth of the regime of research (the Age of Research indeed!) with the extraordinary leap forward taken, at precisely the same period, by socialisation: by socialisation as it came in sight, as I described before, of its reversal point into another hemisphere. There can be no doubt about it: it is not by chance that the number of research students and their interconnections are increasing "exponentially" within a mankind that is in process of concentration upon itself. If you go back to their roots, you will find that the two phenomena are closely allied: or rather, they are one and the same phenomenon, in the sense that

[1] Teilhard uses the English word.

research is in very truth (to repeat more forcibly my earlier assertion) the native and natural form assumed by human energy at the critical moment of release.

This explains how it is that as the unification of the earth progresses, an ever denser and more active atmosphere of inventive and creative interests forms around it: at first you would take it for an insubstantial vapour wafted by every breath of whim or fancy—but in fact it is a formidable, irresistible medium from the moment when, caught up and spun round in the whirlwind of a powerful aspiration it begins (as we can now see with our own eyes) to convolute upon itself, and so attack the real as one single spearhead, following one single concerted direction, seeking not simply to enjoy more or to know more, but to be more.[1]

B. Rebound of evolution and neo-cerebralisation

Evolution makes a fresh start. Still deceived by the slowness of movements that embrace the whole cosmos, we all to some degree find extreme difficulty in thinking of man as still moving along his evolutionary trajectory. We still attribute to ourselves the fixity that we now recognise as an illusion when attributed to the stars, to mountains and to life's long past. Even were it proved that in the course of history, mankind, under the influence of civilisation, has still for some time maintained its way—nevertheless, at the present moment, at the level of individuation we have at last attained, surely, we feel, we must consider it as having finally come to a halt?

With the question so formulated, we reach the point in

[1] In this propulsive system, artistic research, we should note, even though the lines it follows (or a physiology of it) are still obscure and would call for a separate investigation, is not biologically separable from scientific research (which is the only form we are explicitly studying here) and constitutes an integral part of the same exuberant surge of human energy.

this exposition, if I am not mistaken, when we must distinctly and once and for all finish with the legend that continually crops up again of an earth that has, in man and with the man we now see, reached the limit of its biological potentialities. This we can do by showing (still without leaving the plane of scientific observation) that, through the very operation of the forces of convergence developed in the course of a socialisation that is "compressive" in type, the evolution of life on earth does more than simply find a way of prolonging itself in us along the line of its earlier expression: like one of those multiple-stage rockets, it is now visibly starting a fresh forward leap, with a directive mechanism and a power of penetration that are both fundamentally new.

This is a capital point that we must try fully to appreciate. To do so, we should pause awhile and try to form a general picture of the successive steps in the corpuscular arrangement, as the latter appears historically to have been established within a universe in process of involution.

During a first and immensely long period (pre-life) chance alone, so far as we can judge, seems to have governed the formation of the first complexes. Above this (pre-human life) there stretches a wide, disputed, area in which, according to some (the neo-Darwinians), the weaving of the biosphere is again to be explained by chance alone (automatically selected chances); according to others (the neo-Lamarckians) still by chance, but in this case chance seized and used by a principle of internal self-organisation. Higher still (once the threshold of reflection has been crossed) the psychic power of combination finally emerges in the individual among the effects of large numbers as a specific and normal factor of hominised life. And it is at this point, some would maintain, that the biological genesis of invention finally comes to a halt.

Surely, however, it is abundantly evident from what we have pointed out earlier that the cycle is not finished: it is tending, on the contrary, to prolong itself (or rather culminate) in a further terminal state. After "individual" invention, which is the fruit of a tentative search carried out in isolation, we have collective invention, born of totalised research.

And thus we find at the same time that we have reached the heart of our problem. For once we have accepted the relationship noted above between planetary compression, release of free human energy and, finally, the rise of research, we must ultimately reach this conclusion: that a mankind subjected to a compresssive socialisation is synonymous with a mankind braced together in an effort *to discover*. And what does it seek to discover if not, ultimately, the means to super- or at least ultra-hominise itself?[1]

Supposing, however, that we examine what is now going on around us, looking at it from the two-fold angle of continually greater intensification and continually more exact orientation of the effort to discover. We have the physics of the atom, the chemistry of proteins, the biology of genes and viruses—all so many general attacks carefully launched against the sensitive points in which lie hidden the motive forces of cosmic evolution, studied here at its principal levels of articulation—and so many advances, accordingly, towards our conquest of the hidden laws of biogenesis. Until man we have arrangements that come into contact with one another more or less "ready-made" or that tentatively grope their way towards one another in the biosphere. After man (the ultimate and supreme product of this *first-type* evolution), we have arrangements that work themselves out, add themselves to one another, and combine together in the noosphere. Here, indeed, we have evolution mobilising its forces in an effort of a completely new type, made possible by its own conscious-

[1] "Ultra-hominise," on the analogy of "ultra-violet," used simply to express the idea of a *human* prolonged beyond itself in a better organised, more "adult" form than that with which we are familiar.

ness of itself: a *second-type* (reflective) evolution, or (as I was saying a little earlier) the second-stage rocket starting up again with, for zero, the speed built up for it by the first —and, what is more, aimed (as we still have to see) with impeccable accuracy in the same direction—always in the direction of a higher cerebralisation.

2. *Towards more brain.* Earlier (chapter IV, pp. 91-2) I noted and analysed the mechanism of collective cerebralisation that, for lack of other positively observable anatomical evidence, testified to the persistence, throughout the various periods of history, of the movement of cosmic corpusculisation within a mankind that is in a state of expansion. Under a convergent system, it is logically inevitable, and factually demonstrable beyond any doubt, that the process tends to accelerate and intensify. Here again, lost in the vast sweep and the slowness of the phenomenon, we normally fail to concentrate our full attention on it. And yet, all around us and right under our eyes, a process of great importance is going on. It is favoured by the sudden multiplication of ultra-rapid means of travel and transmission of thought, and consists in the formation of more and more psychic zones or groups. In these the human nuclei are converging their powers of reflection upon one common problem with one common enthusiasm, and so organising themselves into stable functional complexes. In these, surely, it is perfectly legitimate, as a matter of sound biology, to recognise a "grey matter" of mankind.

And it is then that a revolutionary possibility becomes apparent, made feasible precisely by the operation of this social innervation (something never before attempted in nature on such a scale or with such elements): the possibility of a new concerted wave of research into the very intelligence from which it emanates: collective cerebralisation (in a convergent milieu) using the sharp spear-head of its vast power to complete and anatomically improve the brain of each individual.

First, *to complete*. And here I am thinking of those astonishing electronic machines (the starting-point and hope of the young science of cybernetics), by which our mental capacity to calculate and combine is reinforced and multiplied by a process and to a degree that herald as astonishing advances in this direction as those that optical science has already produced for our power of vision.

Secondly, *to improve*: this can be envisaged in two ways—either by connecting up neurones that are already ready to function but have not yet been brought into service (as though held in reserve), in certain already located areas of the brain, where it is simply a matter of arousing them to activity; or —who can say?—by direct (mechanical, chemical or biological) stimulation of new arrangements.

Thus, a new and exceptionally central and direct chain would be formed within the noosphere in process of compression: cerebralisation (the higher effect and parameter of cosmic convolution) closing in upon itself in a process of self-completion; an auto-cerebralisation of mankind becoming the most highly concentrated expression of the reflective rebound of evolution.[1]

Although such views may seem very far-fetched there is

[1] Here the distinction between *soma* and *phren* made earlier (chapter II, p. 48) comes in again; this time with such force as to dominate the question. With the appearance on earth of "compressive socialisation" (in which the important factor is no longer the multiplication of individuals but their ultra-cerebralising arrangement) a new system of biological evolution is in fact introduced. In this, individuals, while still functioning as *links* through their *germen* (prolongation of the phyletic in the human, in the form of hereditary fibres that are still recognisable although more and more tangled), assert themselves, through their *phren*, as constituent elements of the "noospheric brain" (the organ of collective human thought).

nothing, I maintain, at all improbable in them. On the contrary, they are simply views fitted to the scale of the dimensions that science meets wherever it engages a movement that is cosmic in scope. There is no better way of convincing oneself of this than to try (as an insatiable curiosity impels us to do) to extrapolate as far ahead as possible the totalising flood of psycho-technical energies, whose convergent advance (as I hope I have demonstrated) is every day becoming more recognisable in the general advance we are involved in.

4. THE UPPER LIMITS OF SOCIALISATION: HOW TO PICTURE TO OURSELVES THE END OF A WORLD

So far, then, from having reached his ceiling (as we are too often told), or even slipping back, man is at the moment still advancing with full vigour: moreover, provided (see below) the planetary reserves of every order do not some time fail him, the ultra-hominising movement now in process—self-maintained or even, as we see it now, self-accelerated—appears to be immune (at least in its most essential part) to the normal threats of senescence. Now that our planet has reached its present level, no physical or psychic force seems capable of preventing man, for millions of years still, from seeking, inventing and creating in every direction.[1] The question, then, is how are we to envisage the general forms of arrangement and consciousness towards which such a current is carrying us.

---

[1] The active life of a zoological family or genus is reckoned at fifty million years. Now man (simply from the point of view of taxonomy) is much more than a family or genus, since, in himself alone, he represents a planetary biological "layer." There are reasons for thinking, it is true, that in this layer evolution, precisely in so far as it is making a fresh leap forward, is proceeding with a constantly accelerated rhythm (cf. p. 116, note 2).

The answer is contained in the decisively and definitively *convergent* character we recognised in the compressive phase of civilisation we have just entered. As a result of the general convolution of the *Weltstoff* that is going on in the most intimate depths of our being, we are advancing towards states that we are justified in describing as "more and more *centred*": and this under three aspects and at three stages, *collectively, individually* and *cosmically.*

Let me explain what the words mean in each case.

(a) *Collectively*, in the first place (and from the point of view of scientific observation this is the axial part of the phenomenon) mankind, as we have had more than ample reason to conclude, tends technico-psychically to converge upon itself. There is no need to re-emphasise this fact, for throughout this chapter it is precisely this thesis that we have been developing: on the other hand, it is most important to note that, precisely in virtue of this process of concentration, the growth of the noosphere is necessarily directed toward some *point of maturation*. At the present moment, in our hope and concern for a quasi-indefinite forward prolongation of man's prospects, there is much talk of the possibility of astronautic inter-planetary migration. I would not absolutely deny the physical possibility nor contest the biological importance of such a diffusion of reflective life within the solar system,[1] but I must point out that to the very extent that this sidereal expansion of our race would give man a wider base for action, to that same extent it could not but intensify the forces that throw us together. If we wish to understand the essence of the phenomenon of man, it is, in the last analysis, to this concentration under pressure—a consequence of the

[1] One thing at least is undeniable: sooner or later the *attempt* will be made by man to overstep the limits of the earth. Does he not feel that, if he is to reach his own centre, he must have made his way to the farthest limit of all things?

world's convolution upon itself—that we continually have to come back. This being so, I believe that what must characterise a mankind that, in some millions of years, will be reaching the polar regions of the symbolic hemisphere in which it is concentrating (cf. above, p. 102), is a higher state of collective reflection; this will be expressed not in a continually greater expansion and diversification of our field of affectivity and knowledge, but much rather by a continually more narrowly localised view of the world (*Weltanschauung*). In this sense, we might say, theoretically and ideally speaking, that mankind will come to an end when, having finally *understood*, it has, in a total and final reflection, reduced in it everything to a common idea and a common passion.[1]

(*b*) Secondly, *individually*—and this in spite of so many contrary preconceptions—there is nothing to prevent us from believing that compressive socialisation, that on first consideration seems so grave a threat to our individual originality and liberty, may not be the most powerful means "imagined" by nature to accentuate and carry to its zenith the incommunicable uniqueness of each reflective element. Is it not a matter of daily experience that union, when acting no longer

---

[1] So that, as I have already pointed out elsewhere (1947), hominisation appears to us as enclosed between two critical points of reflection: the one initial and individual—the other terminal and noospheric. It is at this higher point of organico-psychic maturation, in fact, that the process of "indefinite corpusculisation" (cf. chapter I, p. 31) inaugurated in the world by life, comes to a halt and culminates. Astronomy teaches us that, in the direction of the Immense, the higher unit of grouped matter is the galaxy. Similarly, biology tells us, in the direction of complexity, it is the reflective Noosphere that is, it seems, the higher, absolute, unit of arranged matter: provided, of course, that through space and time "noosphere systems" do not happen, by chance, to be formed in the world—a hypothesis that will seem less fantastic if we remember that, since life is under pressure everywhere around us (cf. chapter I, pp. 35-6), there is nothing to prevent the universe from producing (in succession or even simultaneously) several thinking peaks.

(if we may put it so) tangentially, in function alone (as with the insects)—but radially, that is from mind to mind or from heart to heart, does not simply differentiate but also "centrifies"? The deeper we look into this governing condition of scientifically observable being, the more clearly our minds realise the disconcerting and ambiguous situation of modern man, suddenly confronted by the gigantic magnitude of mankind. *A priori*, and making due allowance for an appropriate reaction from our "liberties," we have nothing to fear from the totalisation that may be expected (as I was saying earlier, pp. 103-4) once it proclaims itself, from its general characteristics (the effects, above all, of psychogenesis) as the legitimate sequel to anthropogenesis. Why this is so, we are now beginning to understand. At the term of the "expanding" phase of socialisation that has just come to a close, we had believed that we were to attain the limit of our own selves in a gesture of isolation, in other words through individuation. From now on (that is, since hominisation entered its convergent phase) it is becoming clear that, on the contrary, it is only as a result of synthesis, i.e. through personalisation, that we can preserve the truly sacred core hidden deep within our egoism. The ultimate centre of each one of us is not to be found at the term of an isolated, divergent, trajectory: rather, it coincides with (though it is not lost in) the point of confluence of a human multitude, freely gathered in tension, in reflection and in one common mind, upon itself.

(c) Finally (and however fantastic this prospect may appear) *cosmically*: if indeed, through its thinking portion, vitalised matter converges, we must of necessity conceive, corresponding to the point of noospheric reflection, *some absolute end* of the universe at the pole of the hemisphere whose dome encloses us. As things now stand, modern astronomers have no hesitation in envisaging the existence of a sort of primitive atom in which the entire mass of the sidereal world, if we took it back several thousands of millions of years, would be found to be included. Matching, in a way, this primordial physical unit, is it not odd that if biology is extrapolated to its extreme point (and this time ahead of us) it leads us to an analogous hypothesis: the hypothesis of a universal focus (I have called it Omega), no longer one of physical expansion and exteriorisation, but of psychic interiorisation—and it is in that direction that the terrestrial noosphere[1] in process of concentration (through complexification) seems to be destined, in some millions of years,[2] to reach its term. A remarkable picture indeed—a spindle-shaped universe, closed at each end (to the rear and in front)by two peaks of diametrically opposite character.

Resembling in this respect Lemaître's primitive atom, Omega point, so defined, lies, strictly speaking, outside the scientifically observable process to which it provides the conclusion: the reason being, that to attain Omega (by the very act, indeed, of attaining it) we step outside space and time. At the same time, this transcendence does not prevent it from appearing to our scientific thought as necessarily endowed with certain expressible properties. Reference to these will be introduced by a concluding study of a final question presented to our minds by the astonishing spectacle of the

[1] And so with every noosphere, each at its appropriate time, should any others already exist or be in preparation (cf. above, p. 114).

[2] If we reckon it at the average evolutionary scale admitted for the genera or families of pre-human mammals, the life of so tremendous a zoological group as mankind would have to extend to several tens of millions of years. But here we must proceed with caution. The "genus of man" does not behave on the tree of life as a mere spray of leaves, or a mere branch, but rather as an inflorescence (cf. fig. 5, and p. 80 note); in consequence, its evolutionary duration could be very much shorter than we think. Even though from the state of organic non-arrangement we can still observe in the noosphere, we might reasonably conclude that man is only now, after a million years of existence, emerging from his embryonic phase.

phenomenon of man. This question may be put as follows: "Projected as we are towards a precise objective that lies in the future, what guarantee have we of arriving at our destination?"

5. FINAL REFLECTIONS ON THE HUMAN ADVENTURE: CONDITIONS AND CHANCES OF SUCCESS

If any point emerges with full clarity from all that we have seen so far, it is without doubt the fundamental and complete inability of human plurality[1] to escape the forces that tend organically to concentrate it upon itself: the general forces of cosmic convolution that make themselves more directly and sharply felt (at the zoological and historical level we have attained) under the influence of the "entry into convergence" of the world in which we live. On that point there can be no possible doubt. In virtue of the very structure of the universe we are forced—condemned—to unify ourselves, if we are to become fully alive.

However, does the fact that we are so situated at the heart of things justify us in concluding that the experiment of which we are the objects must necessarily be successful? in concluding, that is, that we can be *certain*, in any event, of in fact one day attaining the unity towards which we find ourselves impelled? In other words, does the universe concentrate itself above as assuredly and infallibly as it "entropises" itself below?

The answer the facts give is "No": by its nature, and in every instance, *synthesis implies risk*. Life is less certain than death. Accordingly, it is one thing for the earth, by its pressure, to force us into the mould of some form of ultra-hominisation[2]

---

[1] Itself an expression of the atomic origin and the corpuscular nature of every living being.

[2] Cf. above, p. 109, note.

—and quite another for the ultra-hominisation to result. For, if the planetary evolution of consciousness is to reach its term, in us and through us, two series or types of conditions are necessary, external and internal respectively: and none of these is absolutely guaranteed by the progress of time.[1]

First, *the external conditions*. By these I mean primarily the manifold reserves (of time, of material, both nutritional and human) that are essential to keep us supplied until the operation is complete. Should the planet become uninhabitable before mankind has reached maturity; should there be a premature lack of bread or essential metals; or, what would be still more serious, an insufficiency, either in quantity or quality, of cerebral matter needed to store, transmit, and increase the sum total of knowledge and aspirations that at any given moment make up the collective germ of the noosphere: should any of those conditions occur, then, there can be no doubt that it would mean the failure of life on earth; and the world's effort fully to centre upon itself could only be attempted again elsewhere at some other point in the heavens.

Next, *the internal conditions*, by which I mean those bound up with the functioning of our liberty. First, a *know-how to do*, sufficiently expert to avoid the various traps and blind alleys (politico-social mechanisation, administrative bottle-necks, over-population, counter-selections), so frequently to be met on the road followed by a vast whole in process of totalisation. Secondly, and most important of all, a *will to do*, strong enough not to retreat before any tedium, any discouragement, or any fear met on the road.

[1] We should note here that, starting from the moment when (as is happening at present) mankind becomes totalised, there can be no longer any question (as there was in earlier periods) of "disappearing civilisations": there can only be fluctuations and emergences within *one* definitely established planetary civilisation; and this latter cannot perish without *ipso facto* the movement of hominisation on earth being permanently halted.

So far as the conditions of the first type are concerned, it does not seem that we have any particular need to fear the possibility of defeat. From the point of view of material resources and time available, life on earth seems to be developing with a sufficiently wide margin (or with a margin sufficiently enlargeable by technical development—here I am thinking of our reserves of physical energy) for no serious danger to threaten us in this direction: the only exception is a temporary one, from the destruction of arable areas. And from the point of view of our cerebral resources it is remarkable to note how, so far, the human elements spring up and relieve one another, in sufficient numbers and at the appropriate time, in order to carry out the ever more varied and specialised tasks involved in man's progress: as though under the reassuring influence of a mysterious noospheric metabolism.

On the other hand, the internal perils that life has to face, arising from the emergence within it of a reflective liberty—an indispensable factor in its evolutionary rebound, but at the same time a dangerous principle of undisciplined emancipation—these perils, at first sight, seem much more menacing and alive. In this field, even so, we should not forget that the higher reflection rises and the more it builds up its strength (as a result of combined reflections), within the human mass, the more too, *as an effect of organised vast numbers*, do the chances of mistakes (both voluntary and involuntary) decrease in the noosphere. Contrary to what one often hears said, a living system (provided we take it to be, as is the case with man, polarised towards a determined point) tends to correct and stabilise its progress to the extent that the two-fold faculty of foresight and choice arises within its elements at the same time as a sharper awareness of the end to be attained. If you have ten experts tackling the same task, there is less danger of their becoming disheartened and going astray in their work than if you have only one. This means that the more the noosphere convolutes upon itself, the greater are its chances of finally centering on itself.

Even if we accept this peculiarly favourable hypothesis, it remains true that a *super-condition* can be distinguished, essential to the maintenance in action and to the pressure of the ever-increasing and ever-fallible sum total of all our "liberties." This is that, keeping pace with self-reflective evolution, the reasons for an appetite for living (i.e. what we have just referred to as the "internal polarisation") grow stronger in the depths of the human soul. This entails the maintenance around us of a cosmic "atmosphere" that grows continually clearer and warmer as we advance farther: clearer, because we can foresee the approach of a way out through which all that is most precious in what we have worked for may escape for ever from the threats of a total death lying ahead of us—and warmer, from the rising radiation of an active focus of unanimisation. Nothing, apparently, can prevent man-the-species from growing still greater (just as man-the-individual —for good . . . or for evil) so long as he preserves in his heart the passion for growth. But no external pressure, either, however powerful, could prevent him from throwing up the sponge, even with an abundance of energy still available, were he, unhappily, to lose interest in, or despair of, the movement that is urging him on.

This leads up, in conclusion, to the proposition that we may express as follows:

"If the pole of psychic convergence towards which matter, as it arranges itself, gravitates, were nothing other than, or nothing more than the totalised, impersonal, and reversible[1] grouping of all the grains of cosmic thought reflected momentarily in one another—then the world's convolution upon

---

[1] "Reversible" in so far as, with no support to rest on ahead of it, it is structurally bound up with a precarious arrangement of particles that are all, by nature, completely and fundamentally liable to disintegrate.

itself would (in self-disgust) discontinue, in exact step with evolution's becoming more clearly aware, as it advanced, of the blind alley it was ending in. Unless it is to be powerless to form the keystone of the noosphere, 'Omega' (see above, p. 116) can only be conceived as the *meeting-point* between a universe that has reached the limit of centration, and another, even deeper, centre—this being the self-subsistent centre and absolutely final principle of irreversibility and personalisation: the one and only true Omega."

And it is at this point, if I am not mistaken, in the science of evolution (so that evolution may show itself capable of functioning in a hominised milieu), that the problem of God comes in—the Prime Mover, Gatherer and Consolidator, ahead of us, of evolution.[1]

Paris, 4 August, 1949

[1] One might say (and it would be a fair summary of the whole content of this book) that every being (every corpuscle) appears figuratively to our experience as an ellipse drawn with two foci of unequal "power": a focus of material arrangement (or complexity), F 1; and a focus of consciousness (or interiority), F 2.

During pre-life, F 1's activity is practically nil (this is the domain of chance). Then it gradually (cf. p. 108) rises up to the thread of life—until the "threshold of reflection," when the balance is reversed. Starting with man, it is F 2 that takes the initiative in the arrangements that bring about the rise of the power of F 1 (rebound of evolution through reflective invention); while at the same time it becomes progressively more sensitive (to the point of turning in and back upon itself) to the continually growing, and finally exclusive, attraction of Omega.

This amounts to saying that everything comes about, in the course of cosmic convolution, as though the superstructure (the psychic) were gradually replacing the infra-structure (the physical) as the consistent portion of the vitalised particles.

INDEX

Africa, 58, 68, 69, 74, 84
Alaska, 84
albumin, 39
America, North and South, 57, 58, 85, 90
amino-acids, 39
amphibians, 44, 51
anaptomorphidea, 57
annelids, 44
anthropogenesis, 5, 7, 14
anthropoid patch, 38, 57-60, 69, 73
anthropoids, 55-7, 58
anthropomorphs, 56
arachnida, 44
arthropods, 7, 44, 45, 46, 50
Asia, 66
assimilation, 6
astronomy, 6, 24-5
Atlantic Ocean, 59
atoms, explosive expansion of, 32
    formation of, 27ff
Aurignacian man, 83
Australopithecines, 68, 69, 74

biochemistry, 60
biology, 6, 86
biosphere, the, 5-6, 7, 37ff, 40-1, 57
Borneo, 59
brain, development of the, 49, 110-11;
    in equidae, 53-4
    structure of, 51

cellular assimilation, 30
cellulose, 39
cephalisation, 48, 50ff, 56, 75
cerebellum, 51
cerebral hemispheres, 51
cerebralisation, 7, 37ff, 47-57, 76
cerebrology, 50-1
China, 59, 66, 72, 90
Chinese, 65
cholesterol, 39
chordates, 44
Choukoutien, 66, 83
civilisation, 85-92
coelenterates, 44, 50

complexification, 33, 37ff, 47ff
complexity, 21, 33, 34, 40, 46, 47
consciousness, expansion of, 34, 63, 98-9
continents, genetic structure of, 5
convergence, 100-7
corpuscles, 22, 30ff, 40
corpusculisation, 21ff, 35, 46, 47, 49-57,
    62
cosmic convolution, 49-57
cosmogenesis, 5
cranium, evolution of, 75-6
crustacea, 44
crystallisation, 34, 45
Curie, Pierre and Marie, 17
cybernetics, 111

democracy, 94
determinism, 32, 35, 45
differentiation, effects of, 88-90
Dryopithecus, 58

echinoderms, 44, 50
economico-technical organisation, 97-8,
    104
Edinger, Tilly, 53
Einstein, Albert, 34
electrical forces, 33
energy, dissipation of, 32, 33
entropy, 32, 33
enzymes, 39
Eocene age, 54-5, 57, 58
equidae, 53-7
ethnic compression, 97
Europe, 57, 58, 74, 84
evolution, a new parameter for, 47-57
    rebound of, 107ff

fauna, first periods of, 43
    genetic structure of, 5
Fayum, 58
foresight, 63
France, 58
fructose, 39

Gaboon, 59

Germany, 58
Gigantopithecus, 66
glucose, 39
God, problem of, 121
gravitational forces, 33
Greece, 90
Greeks, 65
Grousset, R., 90

Heisenberg, 34
Himalayas, 58, 59
hominian fascicle, 71, 76
hominisation, 37, 64, 65–78, 82–3
Homo sapiens, 70, 74–8, 79–121
Homo soloensis, 66, 67, 69
Huxley, Julian, 21
hydrosphere, the, 6

India, 90
Indian Ocean, 58
individuation, 92, 95
Indo-China, 59
infinite, the, 23
insects, 44, 47, 50
instinct, 7
intelligence, 7
invention, 35, 63, 108–9

Java, 66, 68

Kenya, 58, 69

Laplace, Pierre, 34
leading shoot, the, 46–57
Lemaître, 116
lemuroids, 58
life, complexity of, 21
    fundamental tendencies of, 6
    monophyletic, 38
    polyphyletic, 38
    significance of, 18*ff*
    tree of, 37*ff*, 41–57
lineage, 45
lithosphere, the, 6
Loire, 61

Malaysia, 74
mammals, 44, 47, 51, 53
man, appearance of, 61–78
    consciousness in, 48, 63
    phyletic germinative power of, 76–7

power of expansion of, 73–6
    totalisation of, 96–121
mankind, genetic structure of, 5
man's position in the world, 13, 17
Massif Central, 61
materialism, 34
matter, complexification of, 19, 24
    hominised, 14, 26
    vitalisation of, 6, 22, 48
mechanical skills, 104
Mediterranean, the, 58, 84, 90
Meganthropus, 66
Metazoan animals, 6–7, 44
Miocene age, 56, 76
molecular dissymmetry, 30, 39
molecules and living proteins, genesis of,
    28
molluscs, 44
monocellular beings, 42–3, 47
monophyletism, 38*ff*
multicellular organisms, 42, 47
mutation leap, 74

Narbada Valley, 69
natural selection, 34, 35
Neanderthal man, 67, 69, 70
neo-cerebralisation, 107*ff*
Neolithic age, 98
nervous system, 47–9
noosphere, the, 7, 22, 30, 79–121
North Pole, 81, 102

Oceania, 90
Oligocene age, 58
ontogenesis, 28
orthogenesis, effect of, 90–92

Pacific Ocean, 58, 59
Palaeolithic age, 83, 91
palaeontology, 53, 68, 76
Paris, 61
Pascal, Blaise, 23
Peking man, 66, 68, 83
personalisation, 100–4
phren, 48
phylogenesis, 31, 93
phyletisation, 26, 31
physics, 6
    and biology, 17
    and chemistry, 26*ff*
Pithecanthropus erectus, 66–8, 76

Pithecanthropus robustus, 66
planetisation, 79-82
plants, 46, 49
platyrrhine bloc in S. America, 58
Pleistocene age, 84, 88
Pliocene age, 57, 58, 59, 60, 61, 73 *
polypary, 47
polyphyletism, 38*ff*
Pontian mustelids, 72
pre-hominians, 65
primates, 7, 30, 38
    evolution of, 55-7, 58
progressive relays, law of, 45
proteins, 29*ff*
proto-animals, 43
proto-plants, 43
protophyta, 43
protoplasm, 39
protozoa, 43
pseudopodia, 29
psychic order, 62
    reflection, 63
    temperature, 48, 98
psychism, 6-7, 23, 62

Quaternary age, 59, 66, 74, 78, 91, 92
    man, 66

radium, 17-19
reproduction, 6
reptiles, 44, 47, 51
rhinencephalon, 53
Rhodesian man, 69

scale system, 70*ff*
science, history of, 27
    increase of, 98-9
Seine, 61
Siberia, 84
Sinanthropus, 66, 83
social convergence, 40
socialisation, 79-85, 96-121
    of compression, 96-100

soma, 48
South Pole, 81, 102
Spain, 58
species, sense of, 94
Spengler, 86
'spiritual line', the, 34
sponges, 44, 50
stars, 24-5, 32
Steinheim man, 70
Suess, Edouard, 57*n*

Tertiary age, 59
Tethys, 57, 58
tetrapods, 44
threshold of reflection, 35, 61-78
totalisation of man, 96-121
Toynbee, Arnold, 85-6
Trinil man, 66
trochophores, 44, 50

universe, convolution of, 48, 121
    expansion of, 32*ff*
    of Einstein, 34
    of Heisenberg, 34
    of Laplace, 34

vertebrates, 7, 39, 44, 46, 47, 50, 51
Villafranchian age, 61
viruses, 43
vitalisation point, 37*ff*

Weidenreich, Dr., 91
Weltanschauung, 114
Weltstoff, 113
world, dawn of the modern, 59
    end of the, 112-17
    population, 82-5

Yangtze, 58
Yellow River, 83, 84

zoological scale, 67